THE
NEW ITALIAN CINEMA

Fellini's *City of Women*

THE
NEW ITALIAN CINEMA

Studies in Dance and Despair

R. T. WITCOMBE

New York
Oxford University Press
1982

First published in England by
Martin Secker and Warburg Limited, 1982
First published in the United States by
Oxford University Press, Inc., 1982

ISBN 0–19–520381–X

Library of Congress Catalog Number: 81–85136

Printed in Great Britain

To my wife
and
the memory of
Pier Paolo Pasolini

Acknowledgments

The author would like to thank the following for their various forms of help during the preparation of this book:

in Rome:

Michelangelo Antonioni
Jerry Bauer
Marcella Biagini
Bernardo Bertolucci
Eva Bruni
Liliana Cavani
Margaret Carter
Federico Fellini

Marco Ferreri
Judith Green
Elio Petrie
Francesco Rosi
Mrs Soria (British Council)
Rick Tognazzi and
the Director and technicians at the
Centro Sperimentale

. . .

in London:

Roger Baker
Peter Cargin (British Film Institute)
Mike Cox
Tony Crawley
Michael Dibb
Mike Dorrell
Kevin and Vicky Gough-Yates
Stephanie Greig
Maurice Hatton (Mithras Films)
Jeff Hays
Muriel Hayter
Leonard Holdsworth
Roland Lewis
Tym Manley

Paula Mineau
Luciano and Pat Pagan
Eddie Patman
Simon Redfern
Charles Rubinstein (The Other Cinema)
Elizabeth Salter
Margaret Tarratt
Andrew Tudor
Paul Webster (Osiris Films)
Lindsey Witcombe
Ken Wlaschin (British Film Institute)
and the library staff at the British
Film Institute

. . .

in France:

David Overbey (in Paris)
Jean Gili (University of Nice)

. . .

Especial thanks to Twentieth Century-Fox, for providing the jacket photograph, to my wife, who has indexed the book, and to Laura Morris at Secker & Warburg for her patience and encouragement.

Contents

Introduction

This study of Italian cinema in the past two decades aims to continue the story from the point at which Pierre Leprohon's earlier account left off. But where Leprohon chose to survey the widest possible field I have chosen in the main to study those Italian film-makers whose reputations are international, in order that the book shall remain accessible for non-Italian film-goers.

The working method is also somewhat different, since I have chosen to 'pair' film-makers whose artistic affinities, while not necessarily immediately obvious, are nevertheless substantial when their respective themes, styles and attitudes are examined more closely.

Although the book is thus a comparative study, it owes no allegiance to any schools of criticism. The aim has been to write a popular study, using the traditional critical tools – analysis, description, recall and common sense.

The wondrous thing about Italian cinema in the Sixties and Seventies has been that it has survived on its abundant scale, given Italy's parlous economic circumstances during this era. Why is it, one might wonder, that while Italy and Great Britain have been commonly branded as the 'two sick men of Europe', the Italians have managed to maintain a feature film output of between two and three hundred productions per annum? The answer is severalfold: more government interest and intervention – many current Italian films are funded by RAI Television and receive initial showings on RAI's second channel; more Italians still go to the cinema – the April 1977 *Screen International* report found that there were more box office admissions for Italy in the preceding year than for the whole of the rest of Western Europe; and finally, perhaps, the Italian film-makers themselves, who have continued to make their filmic statements against sometimes very high financial odds.

The decline of their cinema, though perhaps inevitable, has been resisted where possible, and often arrested. Indeed, although the days of neo-realist enthusiasm have long since receded, it is possible to make a case, I believe, for a brief but heady renaissance of Italian cinema in the mid-1970s, to which certain sections in the text attest in some detail.

It is now apparent, in the early Eighties, that all this is changing. Italian cinema faces a threat from pay television channels. Italy's major cities are now being inundated with a plethora of foreign movies. Far fewer Italian films of high artistic calibre are achieving first-run release in Italy's own cinemas. Major directors, like Marco Bellochio, are hard put to achieve any significant distribution for their films even in their own country, let alone in Western Europe or the United States.

The present situation for Italian film-makers, while chastening, is not without odd gleams of light. 'What is needed is a European cinema that joins together the efforts of all the valid cinema forces in Europe,' argues Carmine Cianfarani, President of ANICA, the joint film industry association, in the proposal for film industry reforms of 1978. The Italians already make films in co-production with the French, German and Spanish industries, even, occasionally, with Great Britain. This impetus is likely to be encouraged by the state funding organizations in Italy. European unity is the major theme. Not just co-productions but 'an era of reciprocal collaboration' is stressed by Mario Bregni, Director General of Produzioni Atlas Consorziate. The presence in America of producers such as Alberto Grimaldi and Dino De Laurentis, the latter already importing Italian film talents to Hollywood for his new productions, may well help set the pace. The contexts may prove novel, the indigenous industry may be waning (perhaps directors like Rosi and Fellini, who have always refused to work in the USA, may be the last of their breed), but the Italians have never in the past proved lacking in enterprise when faced with exporting their national gifts. Lina Wertmuller has already done so; Franco Brusati looks about to, and others will follow. Inventiveness and artistry always find channels for expression, even if sometimes these might resemble a last ditch.

1. Exile and Exasperation: Michelangelo Antonioni and Marco Ferreri

There are two distinct phases to the eight films (seven feature and one documentary) that Michelangelo Antonioni has made since 1960. There is the great tetralogy about sexual, emotional and intellectual love, *L'Avventura*, *La Notte*, *L'Eclisse* and *Il Deserto Rosso*, made between 1960–5 in Italy and featuring Monica Vitti, and the trilogy of feature films made outside Italy, *Blow-up* (London 1967), *Zabriskie Point* (USA 1969) and *The Passenger* (made internationally in 1974/5). Finally, there is the full-length RAI TV documentary, *Cina: Chung Kuo*, of 1973, made entirely on location in the People's Republic of China. In this section on the later work of Antonioni we intend to look briefly at the tetralogy, to remind ourselves of its general scope and intentions, and then in more detail at the films Antonioni has made since then.

'Like Ingmar Bergman, Antonioni is a woman's director,' noted Pierre Leprohon in his monograph on the director in 1963. 'It is she who interests him more than anything else, she and her mysteries, her social evolution. It is she who nearly always determines the action or gives it its direction.' This still seems valid as a perspective on the romantic tetralogy, but particularly so of the first three in the sequence, the third, *L'Eclisse*, being the last which would have been available to Leprohon at the time he made this observation. There is also a specific production parallel between the two directors and their relationships, in cinema, with their compelling anima figures, Monica Vitti and Liv Ullman. The priority of attention Antonioni directed towards his heroines was beginning to falter, however, in *Il Deserto Rosso*. There was a near-matriarchal hidden order to the first three films of the tetralogy. As Dominique Fernandez (*La Nouvelle Française*, November 1960) commented, 'Within contemporary society, and particularly in Italian society, the roles have been reversed. The man has abdicated from all responsibility, from all initiative, and it is the woman who has to choose, both for herself and him.'

The male 'abdication of responsibility' in *L'Avventura* is plain enough. In the final shot, when Claudia touches Sandro's head as he weeps in front of her, having betrayed her with another woman, there is grieving strength about her. She forgives him, even though she can no longer admire him. Claudia is a stronger being than Sandro in this moment. Claudia, though,

1

Michelangelo Antonioni on the set of *The Passenger*

become Lydia (played by Jeanne Moreau) in *La Notte*, Vittoria in *L'Eclisse* and Giuliana in *Il Deserto Rosso*, is a woman whose transformations from film to film all lead her towards increasing fragmentation. The breakdown of this underlying matriarchy is, in fact, given our present perspective, the most poignant aspect to the sequence of films today.

One Italian critic, Fabio Carpi, called the main theme of the tetralogy 'the refusal of togetherness'. In a sense, this 'refusal of togetherness' never became complete until Antonioni left Italy. While he portrayed men and women in love or falling out of love, there was always that possibility, however faint, of a true understanding being achieved. When would his lovers walk into the sunset?

In *Il Deserto Rosso*, Giuliana does walk away at the end. But she walks without a man, with her small son, and she walks towards no glowing sunset but a melancholy, industrial horizon. Giuliana is far from reconciled to any man. She does not love her industrialist husband, and she now also knows that she does not love the man she briefly felt she loved, an English colleague (played by Richard Harris). All Antonioni's four heroines have reached the point of no return with men at the end of each story. Giuliana's situation, though, is rather more abject. She no longer likes herself, which is finally of course more important. Giuliana, in the smog, is bereft of her femininity.

The importance of viewing all four films as a sequence is that each film intensifies the process of disquiet. The inner strength of Claudia, which enables her to comfort the weeping Sandro, has suffered a severe fracture by the time we reach *Il Deserto Rosso*, made five years later. Antonioni, in *L'Eclisse* and *Il Deserto Rosso*, moves closer to a criticism of the inherent loss of vitality of middle-class Italian capitalism. Pietro is a stockbroker, Giuliana's husband and their circle of friends are industrial executives. The faintly *déclassé*, bohemian configuration of the earlier couples has vanished. Fernandez asserts that Antonioni is less interested in individuals than in 'the simultaneity of different states of the soul'. This is partially valid, in that the four films, and four couples, *do* interplay as variations on the theme of emotional attrition. But it should also be said that behind these symptoms of loss of vitality is at least one clear cause – the compartmentalism and corporatism of modern life. And, in an *inferential* form, this critique of a form of society which fragments men and women is profoundly political in its objective.

Antonioni is, in fact, one of the most dispassionately political film-makers in Italy. It is for this reason, as well as for the common strand of 'exile', that we have chosen to portray his film-world alongside that of Marco Ferreri's. Antonioni once remarked that he owned no land and owed nobody any obligations. It would be a mistake to treat his self-chosen detachment from the conventional processes of Italian middle-class life – the talk, the food, the sexual banter which Fellini, say, characterizes so raucously – as a form of defeatism. Antonioni believes in a different kind of society, offering a different set of emotional and intellectual options. Since he sees no sign of corporate societies, whether capitalist or Marxist, offering anything remotely resembling what he feels to be a more human way to live, he deliberately chooses to abstain. Unlike his heroes and heroines, he refuses to accept the price that must be paid for material comfort, which is often emotional atrophy. Having documented very fully the various facets of this fragmentation he moves on, outside Italy, in order to intensify his sense of detachment from the wrong social goals as well as revivify his personal creativity.

Since *Il Deserto Rosso*, Antonioni did not make a film in Italy until late 1979, when he began a production in Florence and Rome, once again in collaboration with Monica Vitti – *Il Mistero di Oberwald*, a video-tele-film for RAI based on a theatre play by Jean Cocteau. He has evinced his usual fascination with colour, commenting 'one can take the graininess out of an image and ... make an ambiguous character purple' (*Sight and Sound*, Winter 1979/80).

In the shift from the breakdown of relations (the tetralogy) to the breakdown of the individual (the trilogy) Antonioni's cinema has come to embrace very different ideas about pace, narrative stance and technique. With the

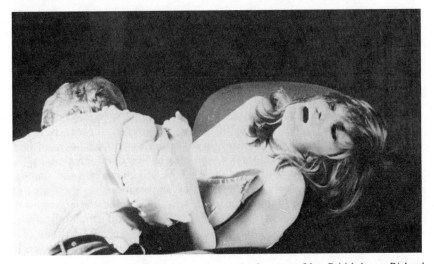

Monica Vitti offering one final despairing orgasm in the arms of her British lover, Richard Harris. *The Red Desert*

crucial shift from black and white to colour (which occurred with *Il Deserto Rosso*) Antonioni became increasingly interested in film technology. The most striking instance of this interest coalescing with visual inspiration occurred in the final seven-minute take in *The Passenger*. Today, still, Antonioni remains absorbed with the crucial technological changes which could help him to extend the purely visual facets of his cinema, or even, as in the case of his current enthusiasm for video shooting, enable him to continue making movies in a world climate which is increasingly hostile to his kind of cinema.

It is this change of dynamic in Antonioni's cinema which constitutes the principal element of the discussion which follows, on the three features and one documentary made outside Italy. The shift, which could be termed a change of gear in Antonioni's own energy, takes him from the penetrating melancholia of the love films to the ebullient exasperation so evident in the trilogy.

But before we consider this shift in dynamics let us mention Antonioni's great missing film, the film he personally still believes would have been his best, which was due to be filmed in 1966/7, was advanced as far as full script, location research and preliminary casting (Jack Nicholson and Maria Schneider, later to star in *The Passenger*, were already booked) and then collapsed rapidly when Carlo Ponti decided against the project in favour of *Blow-up*. The title of this missing film is *Tecnicamente Dolce* (*Technically Sweet*). The screenplay of this film, together with a long disquisition on the projected film by Aldo Tassone, and a short preamble by Antonioni himself, was published in Italy by Einaudi in 1976, though unfortunately no English translation yet exists.

Tecnicamente Dolce merits discussion not only because of its intrinsic interest as a fully developed screenplay but also because Antonioni's career has been tracked by frustrations of this order. Apart from *Tecnicamente Dolce*, there exist in print at least two other film stories developed by Antonioni but never fulfilled. These, however, one entitled *The Horizon of Events*, the other untitled, never reached such a detailed stage of script development.

The regularity with which Antonioni has suffered disappointments in his production ambitions since the completion of his tetralogy prompts one to consider the premise that it is all too often a director's most personal and intimately revealing projects which fall by the wayside. If Antonioni had succeeded in accomplishing, say, *Tecnicamente Dolce* and *The Horizon of Events*, how much, if at all, would our verdict on his later development, through from 1967 to the late Seventies, have altered?

First, the storyline: an impromptu holiday leads a journalist to an island off the coast of Italy (Antonioni had planned to shoot this section of the film in Sardinia). Here he becomes involved with a mysterious girl and her boyfriend. The latter is an anthropology student. The situation rapidly becomes volatile. First, because the journalist (known in the script as 'T') and the girl become lovers. Second, because the student (known merely as 'S') is anxious to leave the island which, remote as it might appear, in fact serves too many brutal reminders of the rapacity of materialist society.

Thus, *Tecnicamente Dolce* contains a double-fugue impetus. First, the flight of T. from mainland Italy to Sardinia. Then the exodus of T and the young anthropological student S from Sardinia to the Amazonian jungle, where they are forced to make a crash landing and explore on foot. This second phase of flight is represented in the structure of the script by a series of brief flash-forwards.

All these main bulwarks of the script's development coalesce to show a new toughness in Antonioni's film-world. It is not that women are devalued, rather that men now emerge from inertia, bad faith and passivity. They begin to take control of their destiny. In this particular and important sense, there is a very direct line through from *Tecnicamente Dolce* to *The Passenger*.

If the masculinity of these protagonists, T and S, is more outgoing, as though Antonioni now needs to frame his characters in new landscapes – the mountains, the cork woods, the jungle – a similar degree of emotional expansiveness affects the love scenes. T's relationship with the girl (who is never named) is a delicate exposition, bearing all the introspective sadness of loving conflicts we know well from the earlier films yet at the same time carrying an *articulated* quality that is entirely novel. There is a scene of passion between T and the girl which shows us Antonioni's grasp of sexual confusion at its most sinuous.

The island holds other forms of turbulence for T besides passion. The fishermen blasting mullet out of the water agitate him. He takes a shot at them, firing over their heads in the attempt to drive them away. In scene after scene complex moments of individual, couple and group relations are crystallized. One of the most striking sequences in the Sardinian segment of the script is the journey by boat made by S, T and the girl out to a tiny remote island. Something appears in the sea not far from them. It is not a fin but a periscope, belonging to an American nuclear submarine.

On the tiny island, where friends of S live in a solitary villa, T witnesses strange experiments which the owner of the place, a natural scientist, is conducting with mullet. In a tower he has an aquarium containing hydrophones. The group sits at dusk, listening to the conversation of the mullet. 'We have recorded them at a frequency that isn't theirs, or we couldn't hear them,' explains the scientist.

The sequence on this tiny island is as carefully orchestrated as the famous sequence in *L'Avventura* when the boat party wanders the rocks and hills after the girl has vanished. But the concept for this island sequence is more richly arranged, allowing for more human manoeuvres. T, for example, is isolated at one stage from the group and walks alone, feeling the gun in his back pocket. This gun which, according to Aldo Tassone, 'annuls time and space', can, equally plausibly, be seen as T's comforter, the sign of his wounded state of mind.

'To travel is also a very sad experience,' Antonioni has noted, 'because ... when you return you are always nostalgic, which derives largely from your dissatisfaction at living in your own ambiance.' The girlfriend of Vittoria, in *L'Eclisse*, clearly illustrates this condition, flaunting her old photos of the African savannah, no longer connected to the past, but equally severed from contemporary reality in Milan. Antonioni, although he made exile his own preoccupation through the years of film-making from 1966 on, has no illusions about its meaning, unlike his protagonists in *Technically Sweet*, who settle for a degree of agitation their creator would find intolerable. T is his mouthpiece here: 'You seek the influence of the forces of nature,' he tells the young people, 'But I want to be conditioned by nothing.'

All the same, T goes along for the ride over the Amazon. And, when the plane is forced to make an emergency landing, finds himself alongside S in fragile conditions.

In the end, both men die, conquered by conditions which are too much for them. But the dying is part of a decision they have made. T dies in the jungle, S crawls to the outskirts, to a ring of hovels where native children play at trapping with a large net. This human animal drags itself, exhausted, into the rim of the net. While the children watch, and one boy messes about with a small wireless transmitter, S expires. His final vision is a glimpse of the little red house above the corkwoods of Sardinia. 'Now,' reports Antonioni with characteristic terseness, 'he looks like an old corpse that not even an animal would eat.'

Tecnicamente Dolce, Antonioni comments in his concluding paragraphs by way of introducing the screenplay to us, 'was a film I absolutely wanted to make ... the drafting was difficult ... because it was developed in the depths of the jungle over almost two months. I saw nothing but jungle and virgin forest.' There he met certain aspects of his creativity in a new way, setting himself those arduous technological problems which he relishes.

However, he adds, 'In the summer of '66 I had overcome all the problems. In the battle between the film and me I considered myself the winner ... when Carlo Ponti said that he no longer intended to produce the film, the world I'd created in my mind, authentic and fantastic, beautiful and mysterious, collapsed. The mortification is still there, inside me somewhere.'

Although Antonioni alludes to himself as the 'black sheep' of neo-realism, knowing that he has long since rejected any solution arising from direct political activity ('to write on politics is to remain in a situation of passive complicity with events,' he once remarked astutely), in one significant sense he remains the most authentic of those Italian *cinéastes* who are heirs to that

David Hemmings and Vanessa Redgrave in *Blow-up*

post-war idealism. He works out of events which impinge on him from daily life. *L'Avventura* was inspired by a boat trip to the isle of Bonza, from which a girl had disappeared, as in his movie. *L'Eclisse* was stimulated by an eclipse which took place in Florence, during which the city came to a stop. *Tecnicamente Dolce* was evolved from his thoughts on the themes of heat, sun, jungle. Its preliminary title was, in fact, *Jungle*. The change of title, however, came from a comment of Robert Oppenheimer's which greatly struck him. Oppenheimer, talking to fellow physicists about the construction of the first atomic bomb, said, 'My opinion is that if one has a glimpse of something that seems technically sweet, one attacks this thing and achieves it.' To Antonioni, this idea was 'irresistible'.

For Antonioni, however, the return to nature is no illusion but a vital imperative for modern man, the means by which regeneration becomes possible. What remains complex, in his domain, is what exactly a 'return to nature' implies. For this most sophisticated of Italian film-makers, 'nature' contains a wealth of implications beyond mere landscape. It is the bedrock states of mind which are released by conditions of physical and mental adversity which he probes in the trilogy of exile with increasing finesse. From the wanderer, T, to the wanderer, David Locke, in *The Passenger*, there is a trajectory more crisply delineated than any Antonioni has made since the bleak meanderings of *Il Grido*.

In 1967 Antonioni shot *Blow-up* in London. It is a film about a photograph, which is taken by a London fashion photographer in a small, somewhat eerie park near Greenwich. Once this photograph is taken, the deliberately diffuse fragments of life so far presented to us begin to adhere around the image. Thomas (David Hemmings), the photographer, is assembling a number of 'documentary' images to make a book collection. When we first see him he is sneaking away from a dosshouse and into his Rolls convertible with his cameras. Thomas, a product of mid-sixties affluence, attempts to regain his dignity by snapping tramps, dossers, vagrants and social cast-outs. For a living, he snaps lean girls wearing *haute couture*, an activity which Antonioni depicts as a form of sadism.

Against this background of professionalized vacuity this picture Thomas has garnered in the windy park takes on a potential meaning. A girl (Vanessa Redgrave) walks with an older man who might be her lover. Thomas spies on them from the fringe of bushes, pressing the button. As they embrace the girl looks towards the opposite rim of bushes from Thomas' vantage point. Later, in his studio, when Thomas develops the roll, he is spellbound by two details in the girl's face: her curious expression, which might be fear or alarm, and the direction in which she is looking. He blows up the tiny frame area showing the patch of undergrowth. There is something wrong. He makes a further magnification. A man is hiding in the bushes. In his hand is a gun. Thomas returns to the park at night. And finds a man's corpse under the tree where he photographed the lovers quarrelling.

At last something momentous is happening to Thomas. He has inadvertently witnessed and photographed a murder. He rushes home to launch this act of genius upon the world he knows, which lives for sensation.

Blow-up is indeed a study of a moral vacuum as serenely sardonic as Pasolini's *Salo*. Thomas, like the 'masters' of the hall of orgies, is a man in the grip of irreconcilable tensions and appetites. He wishes to impose his ego on a world avid for the tastes and gratifications of similar appetites; he oppresses his models, who do not protest because they expect nothing else. He tries to bludgeon into his drugged agent's head the sheer momentousness of his feat and, finally, he returns alone to the site of the killing to find the body gone. Now he is really confronted by his moment of adversity, the bedrock condition to which Antonioni must impel his characters in order to breathe into them the life of authentic awareness, if only ephemerally.

Thomas runs from the park. It is early morning. A car goes by on a park lane, loaded with students wearing the regalia of rag week. Thomas follows on foot. At this moment he is without his sustaining props, the Rolls, the cameras, his sense of self-confidence. He is reduced to a condition which Antonioni considers appropriate for a man who calls himself an artist. Thomas is uncluttered, and curious. He stands by a tennis court watching the students pantomime tennis. When a non-existent ball is hit over the court into the out-field a girl gestures 'fetch it' to Thomas. He runs, hesitatingly at first and then more rapidly, to retrieve the 'ball' as the camera leaves him there, isolated.

Blow-up is adapted from a short story of the same title by Julio Cortazar (published in the UK in the collection *The End of the Game*, Collins/Harvill Press). In Cortazar's story, the photographer, Michel, by intruding with his camera on the lovers, serves a very different purpose. He releases the man – much younger in his version – from the thrall of the woman. Later, contemplating the blow-up in his room, Michel recalls how the boy ran from the woman when they realized the shot had been taken. He thinks, 'I was happy because the boy had managed to escape ... I was returning him to his precarious paradise ...' That sense of regret for lost innocence never enters the movie. Cortazar's view of the girl, the man and the voyeuristic older man who watches them is closer to a Catholic perception of the implied sexual events. Cortazar's hero sees the woman as an enticer, and the older man as an evil presence, all the more menacing for being enigmatic. The innocence of a youth is the moral pivot of the story. Michel is the man of moral sensitivity who reacts to the sexual event in the park with a fine (even a febrile) degree of discrimination. He is, in addition, an intellectual with an intellectual's fascination with potential moral corruption. Hemmings, in the film, returns no one to paradise and consigns no one to hell. Instead, he uncovers a murder.

For Antonioni's photographer, it is a hidden *physical* world, not a hidden moral world, which is detected by the camera. The voyeur in the bushes appears to have a gun. Thus, Antonioni devises a cinematic equivalent, as best he can, for Cortazar's literary and moralizing tension. Was a youth almost ruined by a woman? asks Cortazar. Was a man murdered? asks Antonioni. It seems a fair transposition of interest.

What Antonioni captures very cunningly in *Blow-up* is the rapacity of Thomas's world, which is disguised by the coloured papers of his studio backdrops, the chiffons and curls of the mechanical girls who work for him. Susan Sontag has argued that societies bludgeoned by the photographic image experience life only indirectly. Vicariousness is a symptom of emotional attrition.

The reality of *Blow-up* is above all a visual not a dialogued reality. Characters no longer struggle to lodge a verbal meaning with one another, as they attempted to (often poignantly) in *Tecnicamente Dolce*. There is no yearning in these functional instruct-and-inform dialogues. The film can be grasped most usefully through its visual surface. Above all, the gestural language of Thomas predominates. If Thomas's agent is presented as a man fuzzed by the false calm induced by marijuana, Thomas comes across as hyperactive, technically fixated, dodging from dosshouse to studio to park to dark-room in a never-ending flight from his real problem – the absence of any authentic concentrative capacity.

When I talked to Antonioni in Rome in February 1979, he expatiated on the theme of a technological vanishing point in *Blow-up*: 'To see what the photographer sees in the park – a man and a woman quarrelling by a tree in the distance – is one thing. But to see what the camera eye sees is another reality altogether. Thomas never knows what his camera is capturing. The eye of his camera is more sensitive than his eye. Also, it never changes mood, whereas when we look at someone or something what we see is subordinate to the mood we are feeling at the moment. The camera eye is very precise . . . who knows what would happen if we continued to enlarge a piece of film. A reality that is completely unknown to us maybe. But how far can this process be taken? Going deeper into ourselves, where do we go, what do we discover? These are the questions behind *Blow-up* as I see it.'

'*Blow-up* was to some extent my neo-realist film,' Antonioni remarked in 1976, 'concerning the relationship between an individual and reality, although there was also the metaphysical component . . . after this film I wished to see things more directly, somewhat in the manner of my first films.' In the sense that *Blow-up* is a film which offers a more direct obloquy on consumerist and so-called 'permissive' values, and thereby can be seen to renew the intellectual stringency so much to the fore in his earlier movies, such as *Le Amiche*, *Blow-up* is indeed neo-realist in the broad sense of the word. The direct relationship to reality he talks of we now scrutinize in his movie of 1969, *Zabriskie Point*, which takes him further into exile, outside Europe for the first time in his career and into the California of Flower Power and campus protest.

Zabriskie Point is a film about the two forms of social dementia dominant in America towards the end of the Sixties. The derangement of capitalism is epitomized in the narrative via the land development tycoon (played by Rod

Taylor). He runs a corporation known as Sunny Dunes Estates and wants to develop a tract of desert into a luxury marina. Daria, the heroine (played by Daria Halprin), is his personal secretary. She is a twenty-year-old with romantic attachments towards marijuana.

'It'd be nice if they could plant pot in our heads,' she tells Mark, the youth she meets on the road. 'We'd all have happy childhoods.' Mark is less enthusiastic, 'Make us forget how terrible it is,' he replies.

Then there are the student radicals of the campus sit-ins, who are shown as somewhat bombastic. Yet Antonioni has some sympathy for the causes of the student radicals. In 1968, while making the film, he commented, 'The student movement in America is different [from that in Europe] because it is less together ... When something happens in Rome, it is happening in Italy ... not here. When something happens in Los Angeles, it doesn't matter for New York ... They get in touch sometimes, but they don't work together. At least, that is what they themselves admit.' (*Sight and Sound*, Winter 1968/9). This view of the factionalism of student politics in the late Sixties is borne out by a recent verdict offered by Charles Schwartz, at that time a lecturer and now a professor of physics at Berkeley, who commented (*The Listener*, 3rd May 1979), 'If there were any orthodox Marxists still left in Berkeley in the late Sixties they really must have been looking in amazement at the motley collection of groups which made up what everyone was calling the Movement: this strange revolutionary army that had already managed to accommodate black militants, campus revolutionaries, acid drop-outs, war resistors, mystical hippies and women liberationists. The number of potential recruits seemed almost infinite at the time ...'

It is this ambiance, the raucous radicalism of the Californian youth groups with very little in common apart from rhetoric, which Antonioni carefully establishes at the start of the film. While students argue the theorems of revolution in centrally heated campus halls under the aggressive chairwomanship of Catherine Cleaver (wife of Eldridge and a cult figure on West Coast campuses at the time), Antonioni shows us Mark, present at this debate, his face a hostile mask of disbelief. In deft parallel action, he then sketches in Daria, the boss's secretary, attempting to get into the building to collect a book she left behind.

Antonioni was berated at the time of *Zabriskie Point*'s release for his failure to make a strong political statement in the film. His hero, Mark, does not identify with the procedural obsessions of the student activists. He quits the meeting, accused of being a bourgeois individualist, of not being able to submit his ego to the revolutionary fervour of the group. He returns to his room, puts his revolver in his boot, hears a news item on the local radio to the effect that the Dean has ordered riot police to clear the campus of a demonstration, and goes down to see for himself. A riot is indeed under way. Mark hides, watching, with his gun poised. A shot explodes. A policeman falls. Was it Mark or another sniper who shot him? As so often, Antonioni treats the moments of dramatic action in an intentionally ambiguous way, in

Mark Frechette, rugged individualist, looking for the stroke of inspired action on the streets of Los Angeles. *Zabriskie Point*

order to indicate that his essential interest lies elsewhere. One sees the same obfuscation in Locke's murder in *The Passenger*, and in the murder Thomas witnesses in *Blow-up*.

Thereafter, Antonioni turns the narrative of *Zabriskie Point* in another direction, redirecting his concern onto the road, as he did once before in *Il Grido*, the most sombre and solitary of his earlier movies.

In a useful assessment of the critical abreaction to *Zabriskie Point* in *Sight and Sound* (Summer 1970), Julian Jebb summarizes the adverse impression the film made on both British and American reviewers. He cites Pauline Kael ('There is not a new idea or a good idea in the entire movie ... *Zabriskie Point* is a disaster') and Richard Cohen ('... Antonioni has offered us his contempt').

What, in fact, Antonioni offered Americans (who were at that time narcissistically involved with their own image as fighting warriors, group-Davids embattled with the Goliath system) was a view of their society rather more basilisk than they could tolerate.

'I think that this film is about what two young people feel,' Antonioni commented during shooting. 'It is an interior film. Of course, a character always has his background.' In other words, Antonioni was insisting on his right to remain true to his own strengths, not hypothetical strengths American intellectuals might have wished to impose on him. Moving into an area where rhetorical enthusiasms held sway he maintained his personal equilibrium. He did not ignore the important element of social background. In fact, he took pains with the script, using first Sam Shepherd and then Fred Gardner, the latter himself involved in the Movement, to dialogue the story he had prepared with Tonino Guerra. Fred Gardner's respectful attitude towards Antonioni is, interestingly enough, at marked variance with the howls of the reviewers.

Gardner himself, in 1970, isolated the essentially hollow nature of the now collapsing Movement, 'The problem is,' he said, 'we've won this big cultural victory that implies no corresponding political victory at all. So the country is full of kids who radiate a vague alienation ... who know that the straight life is a dead end, but who aren't political.' This is, of course, what Antonioni succeeded in suggesting in the film eighteen months before.

What Antonioni insisted that we see in *Zabriskie Point* is a love story set on the West Coast of America. How does this interaction between person and place work in *Zabriskie Point*? First, there is Los Angeles, where frenetic billboards and heated people clash and collide. Above the city is the executive, in his large office. In the streets but not *of* the streets is Mark (Mark Frechette, a hitherto unknown actor discovered by Antonioni). Mark is an individualist. He is also audacious, as we see when he strolls out onto the runway and steals the plane, telling a quizzical mechanic, 'Just a scenic flight. Wanna come?' He is also very angry, as we see when he fires the shot at the policeman, a moment of violence which is later reiterated on the road with Daria, when she is accosted by a traffic policeman and Mark hides, gun levelled at the officer, ready to shoot if necessary.

Mark, who tells Daria that until yesterday he has been driving a forklift truck, is angry but not susceptible to the institutionalization of his anger which is what the Movement constitutes. Daria, the girl on the desert road in the Pontiac whom he buzzes in his stolen aircraft, is a dreamer, a gentle girl who is not sure of her destiny and is therefore searching – and susceptible to heroes.

The buzzing of her car takes place on a straight flat desert road, assuming the aspect of a ritual insect courtship. When she finally pulls off the road a red shirt sails from the aircraft. With this, she signals Mark down. 'Did you steal that thing?' she asks as he walks towards her. 'To get off the ground,' he tells her tersely. One American reviewer (*Newsweek*) described the dialogue in the film as 'atrocious'. Yet, it seems to catch not only the romantic yearning of these peripatetic lovers but also the practical streak so common in young, affluent Americans. As they stand on the edge of the vast gypsum craters in Death Valley, Mark tells Daria that he has been kicked out of college for 'reprogramming the dean's computer'. This sort of fantasy – to be a true hero, as Mark sees it, one has nowadays to be technologically subversive – seems very in line with the fantasies of the era.

Down they go, into the dried river-bed. The love scene follows. Again there are graceful echoes of the alfresco lovemaking in *L'Avventura*. After the climax, and Daria's vision of the sterile valley peopled by other lovers, Mark asks, 'Would you like to go with me?' Daria asks where. Mark shrugs, 'Is that your answer?' They climb up the slope. The friction with the curious traffic cop follows. The moment of danger averted, they go to the plane and paint it. Mark is enthralled with his giant, flying mural. Daria suggests that he leave it and drive with her to Phoenix, where she is due to meet her boss. The time for bold gestures is now over. But Mark says goodbye and flies it back to LA, where he runs into a crowd of policemen, and a hail of bullets on the runway, thus demonstrating with his life the validity of his earlier assertion that the students only talk about violence while the police get on with it.

On the road Daria hears the news-flash about Mark's death. In the most moving moment in the narrative, she stops the car and rocks in a layby, her back to camera, a muted WASP version of Catanian keening. When she turns to face the camera her face has changed. It is a devastating moment. This is how middle-class armchair radicals can, on occasion, be turned into Weathermen, Antonioni implies, catching in the body movement, the dark descending hair, the face of ice, the very moment when political fanaticism occurs.

What follows – Daria's arrival at the desert motel where her boss is engrossed in deals in the conference room – is tense: Daria's conversion from dreamer to Fury is juxtaposed with the bland decor; the baffled snarl of wheeler-dealing behind glass doors and picture windows, makes for a jagged, disquieting atmosphere. The sensation of potential conspiracy and sabotage cannot be annulled. Is Daria there to plant a bomb? Is she exchanging a conspiratorial look with that oriental girl in the wash-room? The one vignette within this sullen, ambiguous sequence which impacts without ambiguity is that of Daria in the grotto, bathing her face in the fountain, weeping at last, amid the water.

During our meeting, I asked Antonioni about the curious ambiguity of these terse scenes. The atmosphere, he commented, was determined by the presence of the real estate dealers: 'they are all very ambiguous, because they never trust one another.' In addition to which, of course, Daria no longer trusts them. As always, however, Antonioni is not concerned with closing on a political or social message. Instead, he shows us Daria driving away, stopping her car out in the desert. What follows is as scrupulously orchestrated as the famous seven-minute camera take which closes *The Passenger*. The order of shots in the finale runs as follows:

Daria stops her car. In the conference room a dealer says, 'In this country is gold.' The hotel explodes. Daria smoothes Mark's red shirt on the seat of

The Death Valley love-in, *Zabriskie Point*

the car and gets out to watch. The building explodes again, repeatedly. These detonations occur in slow motion and the camera positions vary so that we witness the trail of skybound detritus with varying degrees of detail included. A table with parasol floats as though weightless, a wardrobe full of clothes, a freezer spills bloody meat and lobster, and then a bookcase eviscerated of its volumes. This is a spectacular dispersal of that accretion of gratuitous goods which constitutes one reality of consumerism. This *dies irae* seems fitting, and tragic. Daria seems satisfied with her apocalyptic vision. She grins and drives off. The camera pans towards a cactus. Antonioni leaves us with no moral. We are open to his finale in terms of our own emotional disposition, which is as it should be, for his cinema is poetic not didactic. Criticized by those who sought to read a film for its political message, Antonioni's refusal to commit himself to message-mongering has been vindicated by the ephemerality of the campus rhetoric. Laurie Taylor, in his report on campus politics (*The Listener*, 3rd May 1979), noted that, 'If you talk to people around the Berkeley campus today, you will find they hardly remember the Sixties: they seem quietly indifferent to the stories of riot and possible revolution.'

In 1973 Antonioni went to China to shoot a documentary for RAI TV. He was there five weeks, filming on a tighter schedule than he was accustomed to – five scenes a day. When the script of his film, *Cina – Chung Kuo* (*China, the centre of the world*), was published the following year by Einaudi he wrote a preface entitled 'Is it still possible to shoot a documentary?' He began by saying, 'Yet again I promised myself I'd write a diary of my journey and yet again I didn't. Perhaps due to my disorder, the frantic working rhythm ... but perhaps, too, for a deeper reason ... to do with the difficulty for me of developing any definitive idea amid all the continuous movement of what People's China is. To understand China it is necessary to live much longer there – during a debate I reflected that he who spends a month there wants to write a book, he who spends several months wants to write an essay, while the man who has lived there several years would prefer to write nothing.'

Antonioni also noted the contradictions China contained for Western visitors and this is indeed the dominant factor contained in different accounts made by visitors from America and Europe. James Macdonald, writing in *New Society* (25th January 1968), noted, 'Life is a flux of contradictions, which exist everywhere and for all time. The fate of a contradiction is to resolve itself in the victory of one of the opposing forces, which then becomes one of the two sides of another contradiction.'

In considering Antonioni's documentary and the reactions to it it seems worth remembering that Antonioni was in China towards the end of the Cultural Revolution, a phase when the doctrinal fanatics held sway, many dissidents were silenced (either shot or incarcerated), and the major institutions, especially education, underwent partial paralysis. That there were rampant political corruptions at work is now clear enough. Antonioni's film failed to register some of the more glaring forms of inequality. Geoffrey Stern, who was in China in 1976, for example, has since commented on that era (*Guardian*, 24th September 1979) in strongly critical terms. The denunciation of Deng's 'capitalist roaders' would have held more weight, he has noted, if the Gang of Four had not been so obviously prone to the usual bureaucratic indulgences – 'Why were so many party and government officials still entitled to first-class travel, a chauffeur-driven car, a sizeable flat, and a salary offering tax-free material advancement well beyond the expectations of most workers and peasants? More important, why the schools for the children of the elite?' This view of Chinese inequalities is at marked variance with the view promoted in *Cina – Chung Kuo*, the commentary to which declares that there was little discernable evidence of privilege.

Yet Antonioni's film remains of interest, not so much for its accuracy of political observation or perspective, which, as we have shown, has been largely superseded by more closely observed accounts of the nation during the years of ideological 'correctness', as for its highly personal representation of parts of the country hitherto unexposed to the film camera.

In his preface to the published script, Antonioni remarked that, back in Italy, he met with quizzical reactions: 'I was asked, on returning, if the

Chinese authorities had limited my freedom of movement, if they had imposed on me a view of life which accorded with the scheme of their propaganda. A journalist noted that in the film the Chinese were always smiling ... But I don't believe that a documentary would have been closer to reality if it had lacked "organized" scenes. The singing children in the homes, and all the rest of the "representation" are obviously images that the Chinese wished to provide, but are not images forced upon the reality of the country ... the Maoism may be propaganda but it isn't a lie.'

If Antonioni did on occasion register the propagandizing ethos without penetrating beneath it, it should also be said that, at other times, he did discern trends in the country's administration which have been substantiated by other, more factually-orientated observers. For example, the commentary records, and the preface to the script underlines the point, that Chinese bureaucracy seemed locally rather than nationally based. When the team wanted to film a factory area called Huang Pu, the local official and the interpreter/guide debated the matter face to face, without fuss. Antonioni deduced that authority in China is not dependent on written regulations, encapsulated as law. He commented, 'I am convinced that the daily life of the Chinese is conditioned not so much by formal laws as by ideas of justice and injustice and from this derives the great part of their simplicity.'

This salient fact about China, its government not by law but by political dogma, has also of course been emphasized by other China-watchers, notably Arthur Miller in 1979. Miller, however, reacts differently to the phenomenon, asking uneasily, 'Does the feudal class system reassert itself time and again because, in effect, there is no legal sanction to slow it down?' His conclusions about China, coming six years after Antonioni's visit, are much less sanguine. China, he believes, during the years of the Cultural Revolution at least, 'was governed like a children's summer-camp: the loudest mouths and the ... reddest of the Red took turns whipping the people on with slogans and guns.'

The film begins in the Tien An Men Square in Peking, where Mao's palatial 'retreat' was situated. 'For the Chinese this great silent space is the centre of the world,' the commentary reports. 'Here the People's Republic was proclaimed.' As if to counteract at once any impression of awe-struck impressionability, a statement of the film's own purpose is laid alongside this rhetorical introduction: 'We don't pretend to understand China. We wish only to observe this great repertory, the times, the gestures, the habits.'

Shots of the old city wall, courtyards inside a students' hostel and interiors of an obstetric clinic follow. Out in the suburbs we see a factory school, a worker's flat, and a factory where workers labour in earplugs. More views of the Chinese Wall, with statistics and also more intimidating information – the bodies of rebellious slave-workers were mixed with the cement – then on to the Street of Animals, a sacred road where lions, unicorns and other holy beasts are paired in stone along its edges. The Gardens of Ming, where 'no one could enter, and even the emperors had to dismount their horses', and, next door, the Museum with its paintings depicting peasants' revolts during

17

the Ming Dynasties. A quotation from Mao is juxtaposed in the commentary: 'Wherever there is oppression there is resistance and struggle.'

In the agricultural commune on the Chinese/Albanian borders we see the canal system which is so vital to the effectiveness of the national agricultural system, a hospital, six clinics, medical schools, and always the faces of the peasants and the workers. 'We would be ingenuous to think we had come across a rural paradise: life here is a hard daily grind,' the commentary reminds us. Then, the first faintly mutinous shooting decision, in the produce market at Hsitan, where the cameras were hidden among the fruit and vegetables and local bartering was observed discreetly, an activity not forbidden by the authorities but hardly encouraged. 'As well as kindness and intellectual subtlety,' the commentary informs us, 'the Chinese have another capacity: gluttony ... but the important thing is the great agricultural return to the land has enabled the Chinese to conquer the Asian tragedy of malnutrition.' With benefit of hindsight, we might now consider 'alleviate' to be a more suitable term than 'conquer'.

Into the old parts of Peking, and the Gardens of the Prohibited City, closed during the period of the Gang of Four (by Mao's widow, for her own amusement, true Ming-style). Another old sector, with the camera tracking past courtyards where 'life is all turned inward' and it is 'difficult to enter the houses and one has to conquer diffidence and timidity, but is then welcomed'. New Peking, bikes and office blocks, intensive traffic. Here, as elsewhere, drab, dun colours, as though Antonioni has at last come across a natural equivalent for those olive tones to streets he worked so hard to achieve by artifice in *Il Deserto Rosso*.

The next sequence shows the rocky mountain district of Linshien, where the poverty is unmistakeable. Main crops are maize and fruit; peasants work with barrows, some under umbrellas. The familiar visual litany, children playing in the mud, Red slogans on the walls, old men resting, follows, together with some facts about the rural commune. The house and implements of the worker belong to him, the land is state-owned.

In the rural commune of Honan other salient facts about China emerge, in both picture and commentary. The absence of cemeteries in China, for example ('place of burial is a matter of choice and is usually permitted'), and the relative absence of migration among the peasants to the cities ('those that go are chosen from an assembly of the entire village'). We see a revolutionary committee pushing a dogma which is conventional Cultural Revolution rhetoric – 'only he who first makes his ideological work right can do all his other work well' is the message from the young leader of the workers' committee. In another region of Linshien Antonioni attempted to film another free enterprise street market and was warned by his guide that he should desist. Antonioni didn't. 'They say it's only a marginal economy,' reports the commentary, 'done for pin money.'

Then comes one of the most striking sequences in the film, depicting the film crew's exploration of a remote mountain village which they enter without first obtaining clearance to film. On a wall poster they note for the first time evidence of protest, concerning the confinement of an Indian tribe in Chinese territory. More startling, however, is the fact that children run away from the approaching camera. 'The Chinese here have never seen a Westerner,' the commentary tells us. 'It is a hard moment for Western pride: for a quarter of the human race we are only known as strangers who inspire fear.'

Following the film's most beautiful visual sequence, the tour round Suchow's elaborate waterways and gardens (during which a young man is glimpsed concealed amidst the stone nymphs and temples, scribbling, perhaps subversively), we contine in Nanking. Here we are shown the new, clean hospital, where acupuncture and herbal medicines are much in use, and then the bridge over the Yang Tse Kiang, shot through a mist from an impertinent lowish angle. Despite the suitably weighty statistics which accompany these shots (the bridge is six kilometres long and took fifty thousand workers eight years to built) this scene in particular subsequently left Antonioni in bad odour with the Chinese authorities.

From the bridge and the schools of Nanking we move on to China's largest city, Shanghai (ten million inhabitants), suitably fortified by a nugget from Mao – 'the flower turns following the red sun of the revolution'. First stop, the famous address – 108 Wang Tze – where the first committee of the Communist Party was formed. There were twelve men in that committee, and of those only Mao 'resisted the storms', instructs the commentary. Shanghai itself is revealed as a vast shanty town, many junks on the Huang Po, houses built from waste materials. The Italian cameras flout a local regulation, covertly filming a war-boat, and then turn to watch a display of gymnastics, everyone at it, even the babies. On this alarming vista of a nation stretching and puffing the commentary concludes with a quotation from an old Chinese proverb:

'We can draw the skin of a tiger,
Not his bones.
We can draw the face of a man,
Not his heart.'

The film's conclusion on this predictable note, as though Antonioni now acquiesces in the familiar retreat of Westerners before China's age and vastness, brings to mind the useful remark about China once made by Francis Hope, to the effect that 'distance breeds trivialization'. The China of 1974 was still immersed in the fanatical revisionism of the Cultural Revolution and, despite such promotional incursions as the Nixon visit, remained enigmatic. Today, doors are opened, and what is revealed, certainly to the judicious eye, is a nation which has been systematically trivialized during its years of isolation. 'That there should be no independent cultural response to reality,' concluded one of the most thoughtful of recent Western observers, Arthur Miller, 'deprives the people of direction as well as of a sense of themselves. Will the Party now resolve to free the artist?'

What was the reaction to Antonioni's film, the first documentary in his repertoire since *La Gente del Po* during the war years? In China the response was generally hostile. James Potts, writing in *Sight and Sound* (Spring 1979) on the theme of international film language, quotes the attack delivered on the film-maker in the Peking *People's Daily*: 'Shutting his eyes to the large numbers of big modern enterprises, the director concentrated on assembling unconnected scenes of poorly-equipped hand-operated enterprises ... fragmented plots, lonely old people, exhausted draught animals and dilapidated houses ... the film resorts to all manner of trickery to deny the fact that the life of the Chinese people has improved markedly.' The shot of the bridge at Nanking in particular was singled out as offensive – 'the camera was intentionally turned on this magnificent modern bridge from very bad angles in order to make it appear crooked and tottering. A shot of trousers hanging to dry below the bridge was thrown in to mock the scene ... the use of light and colour in the film is likewise ill-intentioned. It is shot mainly in a grey, dim light and chilling tones.'

Antonioni himself, in conversation with Gideon Bachmann in *Film Quarterly* (Vol 28 No 4), defended his personal approach to the China documentary as follows: 'We [in the West] point our cameras at things that surround us, with a certain trust in the interpretative capacities of the viewer.' He reiterated the intention which he had made all too clear in the commentary of the film, 'I concentrated on individuals ... to show the new man rather than political and social structures.' What angered Antonioni particularly, however, was the attempts of Chinese authorities to stop the film being screened in certain European countries (in Greece they were successful).

Cina – Chung Kuo, while presenting a very different outline of China from the documentaries of Joris Ivens, whose lengthy study broke down the area of inquiry into a series of compact 'topic boxes' and eschewed the personalized, poetic approach which is characteristic of Antonioni's cinema, is far from antipathetic to many aspects of improvement in the People's Republic. Indeed, given our present perspective, it now seems possible to see *Cina* as too positive a view, as we have elaborated in detail already. Perhaps the most plausible explanation for this contrary reaction of the dogmatists towards Antonioni's movies can be reached by considering Arthur Miller's speculation about the prevailing cultural aesthetic in the People's Republic: 'The very concept of personalized art sits uneasily with them,' he noted. Could it be that Antonioni's highly personal portrait of their continent, so idiosyncratic and eclectic in what it showed, threatened their sense of what art is precisely because it revealed what is missing in their own perceptual apparatus?

'Alright,' shouts David Locke, central character in Antonioni's *The Passenger* (1974), stranded by his broken Landrover in the baking desert. 'I don't care.' He kicks the infernal machine and poses histrionically, arms raised skyward. Yet it is not alright, and Locke does care. Thus, Antonioni begins his most perfectly realized film of exile on two knotty contradictions. A man at the end of his tether has two options open to him: he can collapse, or seek to find himself anew, through the very exuberance which is bred from exasperation.

Locke is a TV reporter. His area is third world coverage. Currently he is in North Africa, trying to track down guerrillas in order to do a filmed interview. The guerrillas, however, are proving elusive.

'When I got back from *Zabriskie Point*,' Antonioni noted in an interview with Philip Strick (*Sight and Sound*, Winter 1973/4), 'after almost two years of life abroad, having practically made a tour of the world, I felt rootless. I was another person really, and I don't know whether it was good or not, but I had lost something and I had gained something else ... looking back at *Zabriskie Point* and *China*, I now see my role as having been an observer, a reporter, and this feeling is certainly carried forward into *Profession: Reporter*' (the original title of *The Passenger*).

The Passenger crystallizes for Antonioni feelings of exile, and the increased exuberance which is frequently the concomitant of that state. David Locke, a man whose defensive professional skin has snagged, tries to annul his frustration by trading identities with a dead man, named Robertson. They have met at the hotel on the edge of the desert, recognizing some similarity

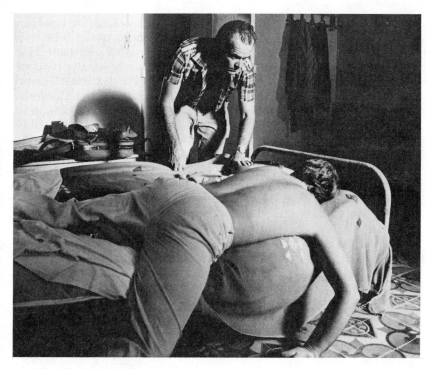

Jack Nicholson in *The Passenger*

21

in one another. Both are lonely men, on the move, deracinated and weary. But there is one telling exchange of thoughts between them, revealed to the audience as Locke dresses himself in the dead Robertson's clothes, via a playback on his tape recorder. Robertson says that he finds everywhere the same. 'No,' Locke replies. 'It's us. We make everything the same. We're so conditioned.' This capacity to try and stand clear of his alienation differentiates him from Robertson. It is his extra shred of vitality.

This theme of making exile real, long present in Antonioni's films, is epitomized most immediately in *The Passenger*. Locke's fugue, to London, Munich, Barcelona and then on the dusty Spanish roads to end in the Hotel de la Gloria, dying a gunrunner's death, is suffused with a strange vivacity. In the festive wedding church in Munich, in the park and on the roof of the Gaudi building in Barcelona, life stirs afresh. He sees the girl, he talks to an old man who has seen it all before, he keeps an assignation with the gunrunners. The details and the significance of such scenes are imprecise by intention. It is part of the director's confidence that he handles the pure plot elements – gunrunners, and the mysterious 'Daisy' – with an aplomb and sense of purpose over and above that demanded by the script, which is somewhat over-insistent in its 'structural' approach to conveying information.

The various formal devices of *The Passenger*, which include time lapse cuts and time lapse within a continuous camera movement, use of voice over playback, use of the camera as a discreet presence in its own right, have particularly interested the reviewers of the movie. Most fulsome of the comments on the 'structural' elements to the film are those of Garrett Stewart (*Sight and Sound*, Winter 1975/6), who calls it 'a project which so effortlessly includes the self-referential interrogation of the film-making process, in its full political implications . . .' What these 'political implications' are Stewart does not explain in any precise sense, beyond the immediate and obvious fact that Locke's interviews with Third World leaders reveal them as far more ruthless than their media image might suggest.

In my view, the political component to the film, involving Locke's own confrontations with Third World leaders as well as the debate about media responsibility which is carried on by his ex-wife, Rachel, and his former current affairs producer, Martin, is less interesting than the personal and poetic aspects to the story. Antonioni, in the three films of exile, hired writers to amplify and supply finishing touches. In *Blow-up* Edward Bond added some dialogue; in *Zabriskie Point* Fred Gardner served a similar function, and in *The Passenger* Peter Wollen's contribution provided a theoretical framework to amplify the basic narrative.

In each of the three films the role of the extra writer serves to provide a contemporaneous feeling to the text, but the *durability* of all the films resides in Antonioni's synthesis of these, the time-locked events and preoccupations, with his own 'timeless' attention to landscape, people and the space between them. Antonioni has absorbed Wollen's 'structural' and Peploe's 'political' sub-texts into his own theme. The political turmoil and the post-Godardian 'metonymics' are redefined to suit Antonioni's own more poetic purpose.

'This is a film about someone who is following his destiny, a man watching reality as reported,' Antonioni himself remarked in an interview with Renée Epstein (*Film Comment*, Vol 11 No 4). 'In the same way I was watching him, in the same way you are pursuing me ... it's surrealistic, isn't it?'

The relationship between Locke-Robertson and the girl points the parallel which exists between *The Passenger* and *Last Tango in Paris*. This parallel is one which Garrett Stewart promotes strenuously in his evaluation. 'In both films,' he says, 'the Schneider character brings body and soul, but no name, to the rescue of an equally anonymous man.' There is, however, a great deal of difference between the two men. Brando, as Paul, is distraught, histrionically malign, given over to gesture. He is on the rebound from his wife's suicide, thus a victim of an equally uncharitable gesture. Locke's metamorphosis, on the other hand, is a reaction to mundane not histrionic crisis – enervation – a failed marriage, loss of professional clarity and zeal and too much scheduled flying.

The striking divergence between the two films lies in the handling of Maria

Jack Nicholson and Maria Schneider in *The Passenger*

Schneider, with whom Antonioni expressed delight during the filming: 'Maria Schneider ... makes mistakes every minute,' he told Philip Strick, 'she doesn't know where the camera is: but that's marvellous ... She has the power to concentrate herself within a scene ... her face is so alive, she has weight on the screen and you look at her.' Schneider's performance bears out this perception. In the scenes with Nicholson, she is candid and fresh. It is a mark of the very different personalities of the two directors that she, in roles which at script level are not so very different, should emerge so differently despite that.

The Passenger, in fact, reflects its maker's buoyancy, not least in the much discussed penultimate shot, in which the camera performs an intricate parabola, starting its majestically slow progress in the room where Locke-Robertson lies waiting for his destiny to enclose him, then out through the hotel window and into the dusty square, where random events are registered – an old man strolling, a boy throwing a stone to a dog, Maria Schneider walking away, a driving school car circling. Antonioni has explained in detail, on different occasions, *how* he achieved this scene, using a gyroscopic camera and a thirty-foot-high crane on which the machine was coupled for the seven-minute take. But on the poetic purposes which underlie such a compelling technical feat Antonioni has remained inscrutable.

What we now see relates most intimately to what we have seen in the body of the movie, throughout which old men, boys, dogs and people travelling in circles have consistently been recorded. The sequence is in fact both a *gathering up* of all apparently random elements into a cohesive pattern, and a *dispersal* of Locke's own identity. Andrew Sarris, writing in *The Village Voice*, perhaps came closest to understanding the sequence in Antonioni's own terms when he noted, 'in these last spectacular seven minutes ... we do not see Nicholson in any active context, but we feel his vibrant presence even in his virtual absence.' Antonioni again shows himself most receptive to the flow of life at the moment when his characters vanish from it.

Antonioni wants to represent *participation* of an ultimate order. In this shot *everything* participates in meaning. Locke, dying, is shared out: 'he exists in and *as* the figures of the random vista. Death is the only authentic democratic moment.'

Although Antonioni has preferred not to comment on what the sequence means to him, he has, elsewhere, discussed the momentousness of death in a barren landscape. It seems appropriate, in conclusion, to enlist his own feelings about death, and eternity, even if we sneak these in from another context.

In *Cahiers du Cinéma* (No 290, July/August 1975) Antonioni introduced the publication of his film story, *The Horizon of Events: Notes for a film to be made or not made*, with some comments which crystallize his ideas of exile, transition and clarity, the master-motifs of his later cinema. 'This new film begins with a trip made in a small aeroplane on a day of bad weather, over Italy ...' He goes on to describe the people on board the plane in detail. Suffice it to say that among the passengers are a businessman and his wife, a writer, his mistress, an elderly lawyer and of course, the pilot. Antonioni then describes the fatal plane crash on a spur of desolate land above the sea. What interests him is the reaction of those who see the crash: two shepherds who live nearby, a policeman and a priest. The priest prays for the victims' souls, and departs, villagers come to gawp and go again. Finally only the young policeman is left to guard the wreckage of machine and people.

Another man appears from the woods. The policeman is uneasy. Could he be a relative? The man wanders amid the debris. Looking is not forbidden. Both men are marooned in a limbo. Soon the site will swarm with pressmen, relatives, insurance men, gawpers. Antonioni interpolates: 'Who these rich people were will be known the next day ... but it's not yet the moment to give to the facts the significance they deserve ... the unknown quality is what matters. It's not for nothing that mathematicians, used to calling things by name, use the designation "x" which is unknown. If I try to determine the "x" of this film I'm led to concentrate on the man who came last upon the scene ... what strikes him most is the sense of a circle, of this implacable horizon. He calls it the horizon of events ... what disconcerts him is that one calls not just the woods, hills and sea the "horizon" but also that line which separates our gravitational world from the black hole.'

Michelangelo Antonioni during the making of his latest film, *Identification of a woman*, scheduled for release in Spring 1982

25

This brief summary of Antonioni's projected film suggests his fundamental curiosity about the nature of phenomena. The film 'to be made or not made' hinges on the thoughts and movements of two men who are suddenly isolated from linear time, social and political purpose, conventional attachments, and are face to face with the awesome moment. Antonioni, writing a film story in which he thinks aloud his most inaccessible ideas, is contributing an explanation by proxy to the ending of *The Passenger*. The hotel-keeper, the couple in the circling car, the boy, the old man and Locke's two women are all caught in that seven minute take in the moment before their 'horizon of events' takes on the potential for expansion. This occurs in the final scene, in which both Schneider and Rachel, Locke's ex-wife, bend over the bed to look at Locke's body. 'I never knew him,' says Rachel, turning away. She has followed him across the world, avid for factual information – is her ex-husband dead or just hiding? He *is* dead. Her interest in the matter is finished. But the girl, in the moment when she says, 'I did' (know him), accepts a different horizon.

The factor which links Antonioni and Ferreri, beyond the theme of exile from Italy which is common to both their careers, is their mutual absence of nostalgia. Neither has ever made period or historical films. Both look at the contemporary world, and portray men in states of acute anxiety and tension, often exaggerating the intensity of these states of mind by means of their highly gestural stories. Ferreri's career as a director began in Spain, prior to which he had edited a film magazine in Rome and worked as a production manager. His first feature film, *Il Pisito*, made in 1958, was adapted with Rafael Azcona from a novel by that author. This was the start of an extended collaboration – fourteen of Ferreri's eighteen films have been co-written with Azcona. The story concerns a young couple in Madrid, desperate for accommodation, who concoct a ruse to achieve this end – the young man marries a wealthy old woman expecting to inherit her estate. The following year, Ferreri made *Los Chicos*, which deals with a squabblesome, over-eroticized gang of layabouts in Madrid, much the same territory explored by various contemporaries of Ferreri's in Italian cinema around this time. One thinks of Fellini's *I Vitelloni*, Wertmuller's *I Basilisci* and Pasolini's *Accattone*, all of which depict high sexual and low industrial energy among suburban layabouts. The third of the so-called Spanish 'trilogy' of Ferreri's was *Il Cochecito* (the only one of the three to be made available in the UK during the early Sixties), which concerned the tyrannical attempts of an elderly man to become more independent of his family by acquiring a wheelchair, in which he contrives to escape with an old friend onto the streets of the city.

The three films, with their emphasis on verbal loquacity and somewhat frantic *mise en scène*, employ many panning and tracking shots, make use of multiple voice tracks, and seem a long way from the Ferrerian world his international audience is familiar with. Later Ferreri films depict a more Spartacan universe, in which images of great ferocity predominate, and dialogue is kept to a minimum. However, assessing the Spanish trilogy in 1961, Enrique Ibanez praised Ferreri's films for their 'humour, poetry, cruelty and love, through which Ferreri acquires universality' (*Otro Cine*, No 50).

Although Ferreri claims that he worked in Spain because he happened to meet and work well with Azcona, added to which production facilities were available there, there is little doubt that the influence of Spanish culture, if not Spanish cinema itself, has impinged on him. 'Ferreri is the only Italian director who might be compared to Bunuel,' observed Alberto Cattini (*Cinéaste*, Vol XI No 2). And, in Maurizio Grande's monograph on Ferreri in the *Nuova Italia* series, Ferreri himself quotes Bunuel's remark that the average Spaniard believes in the dictum 'I have fear of hell but I'm good enough to spit on God', adding with approval, 'Here indeed is a Spanish truth.'

Today, still, Ferreri takes an active interest in Spanish cinema, and has recently promoted the career of a young film-maker, Bigas Luna, with whose film *Bilbao* Ferreri has declared himself impressed. 'When I saw it,' he commented in *Corriere della Sera* (17th February 1979), 'it provoked all the sensations I had experienced when working in Spain, which, at that time, I could not express. It is a very Spanish film, given over to the typical fantasies of the culture.'

What are those 'typical fantasies'? The film concerns a young man's obsession with a prostitute whom he finally takes back to his room, tortures and murders. 'Insolence, repugnance and misogyny,' the three qualities particulary evident in Ferreri's cinema, according to Giuseppe Peruzzi (*Cinema Nuovo* 224, July/August 1973), are obviously dominant preoccupations with Bigas Luna also.

Through the Sixties Ferreri's output was prolific (sometimes two films a year) but the results were uneven and the commercial and artistic impact of the films was intermittent. *Ape Regina* (also known as *La Donna Scimmia*) was banned in Italy in 1963, and the film he made the following year, *L'Uomo dai Cinque Palloni* (*Man with Five Balloons*), was cut from feature-length to twenty-five minutes by its producer, Carlo Ponti, becoming one episode of several in an episode-feature called *Oggi, Domani o Dopodomani*. Ponti's official objection to the full-length film was that it would damage the image of its two stars, Marcello Mastroianni and Catherine Spaak. Oddly, about three years later, at the time Ferreri was completing *Harem*, with Carroll Baker in the main role, Ponti decided to resuscitate the mutilated *Man with Five Balloons*. New scenes were shot and the black and white print was treated to a colour process. The idea was to aim the film outside Italy for exhibition.

Harem (1967) marked a shift in Ferreri's focus. The film began as a black comedy but, during editing, Ferreri decided to alter the dramatic emphasis and play it straight. The story concerns a Milanese architect, Margherita (Carroll Baker), and her relationships with five men, all of whom try to degrade her in different ways. In one scene a plate of spaghetti is thrown in her face.

Marco Ferreri (centre) changing roles and seen here playing one of Pasolini's trio of capitalist villains in *Pigsty*

Finally she is murdered on a beach, a fate shared with other Ferrerian characters.

In Ferreri's next movie, *Dillinger e Morto* (*Dillinger is Dead*), made in 1968, the wife of the hero is murdered. The film depicts the activities of a hypermanic executive (Michel Piccoli) during one night in his gadget-ridden home. He cooks and cleans his revolver. He seduces a maid while his wife is nursing a migraine in her room. He runs home movies, using his torso as a screen. Finally, he shoots his wife in the head and leaves. The film ends with his departure by boat from a remote shore. On the horizon hangs a red ball. Ferreri's 'horizon of events', however, is rather different from Antonioni's, suggesting apocalypse, not a tentative exploration of the threshold of awareness.

Between 1968 and 1972 Ferreri produced Godard's *Vent D'Est* and Rocha's *Lion with Seven Heads*, and he made television documentaries for RAI TV, among them *Perche Pagare per Essere Felice?* (*Why pay to be Happy?*), a study of tribal hippy groupings in the USA. He also directed the feature films *Il Seme dell'Uomo* (*The Seed of Man*), *L'Udienza* (*The Audience*) and *Liza*.

In this period Ferreri's sardonic cast of mind began to alter. Both *Il*

Seme dell'Uomo and *Liza* push him further towards an apocalyptic, vaguely 'future shock' vision of the new barbarism. Only *L'Udienza*, a satirical study of a poor believer's abortive attempts to receive an interview with the Pope, shows signs of the old, mordant, Ferreri/Azcona iconoclasm.

Superficially, there is a resemblance between *Il Seme dell'Uomo* and *Zabriskie Point*. In both films landscape plays a major part. Antonioni uses the gypsum wastes in Death Valley and Ferreri the duney beaches in Tuscany (perhaps at Maremma). The characters are dwellers in the wilderness, in flight from technology and bureaucracy. They commit bizarre acts whose significance (and this is, of course, especially true of Ferreri's film) seems fairly private. Both films, above all, end in the apocalypse of explosion.

We shall discuss *Il Seme dell'Uomo* in detail. First, though, in brief, *Liza*, a film about a man in retreat from his wife and family, who is living on a small island with his dog Melampo. The Pasolinian ambiance of the film's mythic presentation is most evident when the man lets his dog lead him over the rocks, pretending to be a blinded Oedipus.

'I have known this island since I was fifteen,' Ferreri has remarked, 'It's the first time that I derived my impressions from a precise place ... in *Il Seme dell'Uomo* I used landscapes to convey certain ideas I had about the future. In *Liza*, the situation of the characters comes from the island itself.' (*Jeune Cinéma 64*, July/August 1972).

The island drama, developed by screenwriter Jean-Claude Carrière, largely *in situ* at the director's insistence, shows the middle-aged ex-family man (Marcello Mastroianni) contending with a threat from the outside world in the form of a girl (Catherine Deneuve) who arrives off a yacht and rapidly becomes demanding. She kills his dog and puts herself in its place, acting out the role of mute, dependent companion from one corner of the hut, but never looking plausible in her dog collar begging for scraps. As in Wertmuller's *Swept Away*, Ferreri views this encounter between a scared man and a rich woman as primal conflict. There is violence, interspersed with flashes of tenderness, moments when they sit together watching the sea and sharing food. There is also, however, the city which the hero has abandoned, emerging in an abbreviated parallel action which shows the abandoned wife and daughter assuaging their sense of loss by gluttony. This element foreshadows the main theme of *Blow-out*.

In *Il Seme dell'Uomo* we see Ferreri's obsessive imagination at its most florid. Ferreri has denied that he is drawn to an apocalyptic cinema, preferring to substitute the word 'Elsewhere'. 'We have already moved imperceptibly into an "Elsewhere",' he has noted. 'That is, inside a reality that merely survives itself or that is already obsolete' (from his programme notes to *Ciao Maschler* (*Bye Bye Monkey*)). Yet there is without doubt an apocalyptic note to *Il Seme dell'Uomo*, as there was to the ending of *Dillinger e Morto*. On a beach a boy (Marco Margine) and a girl (Anna Wiazemsky) camp out in a stone hut, having fled from a city – in the opening sequence we see them wandering about in an autostrada cafe, and then an airport. This opening sequence suggests a society collapsing and consuming. We see bomb-ruined buildings, and people eating heartily in restaurants. The ruin, we realize, is only a device, projected onto walls for the delectation of the diners. Yet, once the young couple leave the city, social menace is palpable. On a strange helicopter launch-pad an injured man is being attended. 'What's happened?' demand the couple. They are told tersely, 'What should have happened has happened.' They move on, into arid country. 'They're burying the dead,' says the boy. The feeling is akin to Cavani's *Cannibals* – two young people cast adrift in a harshly authoritarian wilderness.

Once on their beach they encounter a fat man asleep in a dune – played with malign relish by Ferreri. As they try to settle in their stone hut shots of cities collapsing are intercut – London is burning, to a thunderous soundtrack of Verdi's *Risorgimento Song*. Soldiers outside St Peter's say they want to speak to Il Papa. But Il Papa, when visualized, is prone. On the beach the boy and girl are burying a dead man and playing a hunting horn. 'Elsewhere' may have come, but there seems little food shortage. They stagger indoors with a huge cheese.

The couple, however, are not alone on the beach. Horsemen arrive, hippy-looking, to interrogate them, then ride away again. Inside the house it seems like a paradigmatic 'marriage'. He has Do-It-Yourself books. She peers through a telescope, or sits by the sea regarding a Mona Lisa print. In the background covered wagons are seen. A gang of men take part in a boar hunt and dismember the carcase. A dead whale is found on the shoreline, on which the couple climb, reminding us of Italy's much-loved folk-novel, *Pinocchio*, in which the boy-hero gets swallowed by a massive cetacean.

With the arrival of Annie Girardot (another Ferreri regular) in a martial-looking combat jacket, the drama gets bloodier. She is a typical Ferrerian older woman, brisk and manipulative – we see this type again in *L'Ultima Donna*, but more marginally. That night the three of them share a bed, Girardot now wearing a long black wig while Anna Wiazemsky pretends to sleep through furtive sounds of passion.

Carnality now cedes to carnage on the beach, with the girl murdering the older woman, cutting up her corpse with an axe, and feeding cooked chunks of her former rival to the young man at supper. Ingested the rival might be, but not forgotten. The young man makes a sand effigy of her and the girl discovers him lying with it. Further hints of an authoritarian outside world impinge when officials arrive to drive the couple from their cabin in order to fumigate it. More fumigators spray the beach. The lovers take refuge under a waterfall, he naked, she in a boilersuit, that ironic inversion of male/female stereotypes which Ferreri often finds amusing.

Now Ferreri combines his recurring symbols – beach-monsters and manni-kins – in one surrealistic scene. The young man talks crazily to a group of propped mannikins in a warehouse. 'I'm not going to take you away,' he reassures the effigies, 'because you're all in couples.' The girl comes in to tell him that the whale's been eaten. She does not share his tender feeling towards the mannikins. 'I don't want a graveyard of dolls,' she tells him. Natural sound, a strong element in the film – wind, seabirds and waves – is unnaturally heightened. Images of bizarre *dulce domum* follow. He sits at his microscope; she basks in a deckchair, holding a white dove. That night, while she sleeps, he writes in his journal, 'Today will be sown the seed of man, I hope.' He sits by her on the bed, fumbling with her thighs in a ceremony which is intended to mystify us. Artificial impregnation? The question lingers through the final shots. She walks on the beach under a green parasol while he, in a loin-cloth, is raking the seabed. 'How do you feel today?' he inquires. 'I've got stomach ache,' she tells him. They sit under the whale skeleton and he tells her she is going to have a baby. She panics, 'Why did you do it? Why?' She covers her face in dread. An explosion puts an end to them.

It is possible to evaluate *Il Seme dell'Uomo* in terms of its own themes of contamination and sterility, and the descent into an enigmatic, post-industrial barbarism. It might be more useful, however, to consider Ferreri's movie in relation to his later ones. Ferreri's main compulsions are discernible in the film (all the more so for the conspicuous absence of Azcona's dry humour in the script) but are worked out more coherently later. *Il Seme dell'Uomo*, like Cavani's *Cannibals* and Bertolucci's *Partner*, is a very rhetorical film. All three films, made in the Flower Power era, between 1967–9, carry their own schizoid signals. All reject authoritarianism. Yet all posit an alternative society which seems just as fascist, in which murder and cannibalism are taken for granted.

Il Seme dell'Uomo, however opaque, seeded certain films of Ferreri's which followed in the Seventies; in particular, *La Grande Abbufata* (*Blow-out*), *L'Ultima Donna* (*The Last Woman*), and *Ciao Maschler* (*Bye Bye Monkey*), made in 1973, 1975 and 1977 respectively. Society in its death-throes, gorging, consuming itself to extinction, committing acts of tyranny and acts of self-castration (tyranny's obverse), and finally sinking into infantilism. Ferreri's cinema has remained deeply enmeshed in destructive dreams, which he might call visions. The obsessional nature of his art is most readily grasped by the recurring symbolism. The linking symbols through from *Il Seme dell'Uomo* to *Bye Bye Monkey* are plain enough. Both films make a powerful symbol from the dead beach-monster, in the earlier film a whale and in *Bye Bye Monkey* a gigantic King Kong creature. Both make use of the mannikin as an emblem of disgust, wax creatures which die by melting.

The relationship between *Il Seme dell'Uomo* and *Bye Bye Monkey* is not unlike the one existing between *Tecnicamente Dolce* and *The Passenger*. Both directors, either failing to realize (or realizing imperfectly) a work from the late Sixties, returned to the same motifs in the mid-Seventies and re-organized themes from the earlier works which they must have felt they had not exhausted.

Lafayette in *Bye Bye Monkey* works in the Wax Museum of Ancient Rome in New York City, rides to work on a bike and lives in a dingy basement. One day on the beach by the Hudson river, out with three acquaintances, he finds a giant chimpanzee and a new-born monkey nestling in its carcase. Lafayette calls the animal Cornelius and becomes much attached to it. At the same time he develops a liaison with a girl called Angelica (Gail Lawrence), who belongs to a feminist theatre group where Lafayette helps with the lighting. Angelica becomes pregnant by Lafayette but runs away, claiming he is not capable of becoming a father. At this traumatic moment Lafayette's crisis is deepened by his discovery of Cornelius' mutilated corpse in his basement.

Returning to the Wax Museum, Lafayette tells Flaxman his troubles. But the Museum's narcissistic director, interrupted as he declaims Shakespeare's 'Julius Caesar' in full costume, has no sympathy. Lafayette throws him against a statue. Flaxman collapses against a switch, causing a short-circuit which sets fire to the building. Once again, Ferreri ends on the destruction of wax effigies, except for one telling final shot, depicting Angelica sitting naked in the sun, feeding grapes to her baby daughter. 'While Man is cracking up, Woman is ... growing up,' Ferreri has commented *à propos* this ending.

Bye Bye Monkey, then, transposes *Il Seme dell'Uomo* to New York City, dropping the earlier film's theme of sterility and reducing the theme of contamination to socially plausible proportions – Lafayette, like the young man in *Il Seme dell'Uomo*, is also puzzled by men with fumigation equipment but this time Ferreri explains their presence; they are rodent exterminators.

If the film's imagery relates back to Ferreri's work in the Sixties, the dominant theme, that of woman's inherent evolutionary superiority, is very much part of his Seventies ideology, featured still more emphatically in *Blowout* and *The Last Woman*. One has the curious sense that Ferreri's grand theme in the Seventies echoes Antonioni's in the Sixties: that is, he has sought to explore the possibilities of matriarchy by male default through these three films, as Antonioni did in *L'Avventura*, *La Notte* and *L'Eclisse*. How effectively Ferreri has tackled this important contemporary theme is the question which underpins our examination of *Blow-out* and *The Last Woman*.

Blow-out, described by Maurizio Grande in his monograph on Ferreri (*La Nuova Italia* series, Castori, Italy 1975) as 'a projected dystopia of destruction and decomposition', certainly represents not only Ferreri's dystopic but also his dyspeptic vision. It is a film about gluttony. A group of middle-aged men – a lawyer, an airline pilot, a master-chef and a TV producer – assemble in a Parisian house to eat themselves to death. The surfeit of rich food and drink, however, hardly obscures the film's links with Bunuel's *The Exterminating Angel*. In *Blow-out* we see Ferreri's intentions running at their most Bunuellian pitch. Yet, where Bunuel, in his study of the upper middle-class protagonists under siege in an elegant mansion, retains a sardonic distance from his subject, by use of characteristic surrealist curlicues, Ferreri closes with his deranged gluttons, recording Michel Piccoli's agonized, strident farts as he stands at the piano, the sweat on the advocate's fatty jowls as he struggles through yet another pâté, the chef's desperate tension as he chops, guts and minces, and the pilot's sadistic lust as he bends his blonde whores over the bonnet of a Bugatti in the garage.

Ferreri, with his insistence on the image not the word, the negation not the celebration, here amply justifies his view on cinema – it is the gritty imagery which lingers; Ugo checking carcases as they are brought from the delivery truck parked in the decaying garden – 'six young saltmarsh lambs, twenty dozen chickens;' Philippe (Philippe Noiret), the fat advocate, tippling at blood pudding and telling a quizzical Chinese visitor, 'We are having a gastronomic seminar'; Marcello in the garden stroking the stone statue of a woman with all the gloating avarice of Fellini's Casanova.

Ferreri's least impressive trait is his dependency on private dogma. Juxtaposing his cinema alongside Antonioni's serves to expose this, for Antonioni is the least dogmatic of Italian film-makers. In *Blow-out*, however, Ferreri grapples vigorously with his dogmatic tendency. He separates and inspects the individuality of appetites – Marcello cannot gorge himself without also fornicating, a truly Rabelaisian imperative; Philippe, on the other hand, who likes sweet things best, is initially outraged when the introduction of harlots to the banquet is proposed and swiftly seconded. Michel, the liberal media

Gail Lawrence and Gerard Depardieu as the uncertain lovers in *Ciao Maschler*, Ferreri's first movie with a New York setting

Michel Piccoli, Philippe Noiret, Ugo Tognazzi and Marcello Mastroianni in *Blow-out*

man, allays visceral pressure most preciously, at the barre in a leotard, to a touch of tinkling music. While Ugo, the only member of the quartet hailing from a conventional family background – we see him at home in the opening sequence packing his knives while he tells his wife the weekend away is strictly business – retains a tighter grip on his appetites. While Philippe and Marcello roll with whores on a pile of furs Ugo hangs on a shade grimly to tomorrow morning's menus.

The arrival of the whores intensifies the dementia. They are wined, dined and sullied. But they, with their sensitive antennae for derangement in a client, are never easy at the orgy and depart in the morning. Like rodents, they sense this ship is sinking. Madame Andrea, however, the provocative Parisian schoolteacher who arrives inadvertently on the threshold and accepts an invitation to supper, is very different. Nothing will shift her. She, like the whores, detects morbidity but is fascinated. Andrea Ferréol, in this role, performs with a steely roguishness which only reckless gluttons could fail to find terrifying. As the tribe's appetites foul up on excess, she helps them one by one towards oblivion. She coaxes Marcello into racing up and down the drive in the Bugatti. He freezes to death in a blizzard, in his flying cap. That brief futurist flowering of Italian energy and inventiveness, symbolized by Marinetti and the Bugattis themselves, seems encapsulated in this snow-veiled frozen image.

Marcello is the first to succumb. The survivors stash the corpse in the freezer and stuff themselves faster, with no more semblance of enjoyment. Madame Andrea becomes dominant. Coaxing chef and diners on to death, she becomes a Medean figure, cajoling her boys towards fresh excess, fellating and berating, and serving the poison. Ugo's paternal cry – 'If you don't eat you won't die' – now stiffens into an iron dictum in her hands. The kitchen becomes a nursery. For Philippe (who is still attended at home by an elderly nanny) it is home from home.

Next to go is Michel, staggering from piano to balcony, there to collapse in a maelstrom of intestinal disturbance, mimicking Brando's 'last tango' sans gum and with gut effects. As he dies dogs bay loudly in the garden. Soon there will be meat in abundance, the hounds sense. Indoors, Ugo, in tall chef's hat, announces, with the last shred of culinary authority, 'the final meal will be a *brioche*.' As they eat Philippe asks pettishly, 'why the egg slices?' Ugo tells him, 'eggs are the symbol of death, according to the Jews.' He munches on relentlessly and, as he lies prone, Madame Andrea performs Consumer Man's last rites, masturbating him across the veil, after which she takes Philippe to bed while the dogs howl.

As she spoonfeeds Philippe blancmange the next day, a meat-truck delivers. 'Tell your boss the veal was second-rate,' shouts Philippe to the delivery man, but it's the bombast of a dying man. Andrea pushes his fat corpse away and walks into the mansion through a garden now carcase-draped. The dogs come.

Blow-out is Ferreri's most visually inventive movie, but shows him still grappling with his theme at an essentially rhetorical level. Ferreri's meta-phorical depiction of western consumer societies choking to death on the great glut of goods is, in the end, hollow, for he suggests that the end-game of hyper-consumption is extinction. It is a comforting fantasy, but fails to allow for the resilience of the technological society which, much as artists like Ferreri might wish its extinction, refuses to collapse, and merely develops new survival mechanisms.

Eating as suicide is a mysterious phenomenon. 'Problems of over-eating and under-eating – obesity and anorexia nervosa – have been extensively studied, but with disappointingly little practical success,' observes John Nicholson, a lecturer in psychology (*New Society* 4th April 1976). He adds, 'eating disorders are unusually resistant to treatment: analysts fight shy of patients with these problems.' Gluttony is not only enigmatic, but also, it seems, an unattractive disorder.

Central to the collapse of recognizable social forms, in the Italian context especially, is the collapse of the family, according to Ferreri's diagnosis. 'The concept of the family,' he told me, 'was a concept of force. The family was a tribal tie. Now "family" does not imply a network of relationships but a network of shadows. Distance exists between family members. Instead of a family, you have a telephone.' Ferreri's preoccupation with the attrition of that once robust bastion against state encroachment, the extended family network (particularly potent in Italy until recently, and a sufficiently distinct social phenomenon to be called by sociologists 'sottogoverno'), can be studied clearly in his most consistently coherent movie, *The Last Woman.*

The film concerns a young engineer, Gerard (Gerard Depardieu), who works in an oil refinery and is undergoing considerable domestic pressure. His wife has recently left him, and Gerard is trying to bring up their son, Petey, on his own. If this suggests a somewhat feckless absent mother, it should be said that Gerard exhibits a degree of personal confusion which would make him intolerable to live with. His wife is now living with another woman, and has left Petey with Gerard because, as she tells Gerard's new (very young) girlfriend, he would not be able to cope without Petey around to console him.

Ferreri has said that he was less interested in the man–woman relationships depicted in the film than in the new element he introduces, that of the

Andrea Ferréol and Philippe Noiret during the last phase of the great apocalyptic eat-in – *Blow-out*

toddler: 'The child acts as the catalyst for much of the situation,' he remarked (*Cinema Papers* No 10, October 1976). 'It's the child who causes the protagonists' need for a relationship that is maternal not paternal ... the man seeks such a relationship with his son because he is trying to rediscover himself ... his ideal woman is a madonna.' When I spoke to Ferreri in February 1979 he reiterated this idea still more emphatically, declaring that relationships between men and women are no longer of interest to him – 'we talk of the man and the woman, but we never talk of the child.' If this suggests a degree of social compassion surprising to those who are acquainted with Ferreri's cinema and the themes it embraces, it should be added, I believe, that what it may suggest about Ferreri's drift of thought is a *transfer of abstractions*. Ferreri himself provides the dogma which underlies *The Last Woman*, in the following assertion:

'The couple is an institution external to man; it's the most repressive superstructure ... fear of solitude is at the bottom of it ... the couple serves as a pawn in the economic and power games of society, games which are extraneous to man ... instead of uniting, it fragments the tribe, the clan, the family. A revolution of the system is necessary ... to accustom man to solitude.' (*Cinema Papers* No 10.)

Such a ready propensity to generalize is a marked feature of Ferreri, and in acute contrast, of course, to the more tentative speculations of Antonioni. It can be seen, then, that the parallel attachment evinced by both film-makers for the contemporary world (which for both of them represents Europe and the United States as well as Italy) is very different once the detailed fabric of their suppositions about cinema and society is scrutinized. Let us return to *The Last Woman*, in order to examine how Ferreri's handling of this man/woman/child trinity is developed.

At the heart of Gerard's confusion about his masculine identity is his ambivalence towards women. They are increasingly threatening to him precisely because they now seem less neurotic than him. The opening scene makes this clear. We see him strutting about the oil refinery compound, slanging a foreman, then leaving on his powerful motorbike to collect Petey from the playschool.

It is the girl, Valerie (Ornella Muti), from Petey's playgroup who moves in with him. She is beautiful, perhaps ten years younger than Gerard, and attracts him with her self-possession. Back at the flat, having shrugged off *en route* a puzzled lover of Valerie's (Michel Piccoli) who was expecting to take her away on holiday but accedes gracefully enough – 'You're your own woman' – Gerard demonstrates his versatility by making love to Valerie, with Petey gurgling between them. During these preliminaries to the relationship Valerie pleases Gerard by saying, a bit dryly, 'You're a good mother.' What she does not say, but what she gradually comes to feel as the relationship deteriorates, is, 'But you're not a good father.' Gerard's attachment to his son is too obsessive. He treats the child as a distraction from his melancholia and fragmentation, waking him in the night because he cannot stand solitude.

Much later, when Gerard is showing Valerie old snapshots, she looks at one of him as a child and says sadly, 'I like you better as you were then.' She has understood that his caring capacity is too wayward. Once, when Gerard insists on making love, the child is playing unattended, close to a broken glass. Ferreri low-angles this scene to underline the incipient danger.

For all Gerard's faults, however, there is something touching about his alternating mawkishness and bragging, and undoubtedly Depardieu's empathetic performance helps the film greatly. More than other Ferreri heroes, Gerard is an arrested personality, but not altogether a deteriorated one. At least he is trying to change himself. He retains a measure of vitality, but the light flickers; his control of himself is slipping. He plays obsessively with an electric meat-carver, and once he cuts his finger on it. Gerard is a man attempting unsuccessfully to break free of phallocentricity. 'Who puts the seed in?' he demands petulantly when his ex-wife calls round to check on Petey (and the new girlfriend). 'Oh, there's plenty more where that came from,' she says. Her careless manner adds to his discomposure. Later, he lurks at the high apartment window watching his ex-wife, Valerie and Petey walking in the gardens below. In this image Ferreri catches the essential isolation of Gerard. The ease with which women around him form relationships eludes him. In fact, Ferreri underlines the power of new sorority, which is perhaps the latent theme of the film, via the wife, who wants to entice Valerie away from Gerard into her own all-woman menage.

This is a pertinent detail, suggesting that the wife's possessive feelings have transferred; it is not her son she wants to steal away from Gerard, but the girlfriend. Does this imperialism stem from her desire to protect Valerie, or deprive Gerard? Unfortunately, the wife is too sketchily presented; beneath the casual, 'liberated' surface, she remains an emblematic character.

The most evident failure of intelligence in *The Last Woman* begins and ends with Ferreri's presentation of the women around Gerard. By making their virtues so apparent, alongside Gerard's nervous bullishness, his rough touch with the child, his phonily bantering relationship with a neighbour of his downstairs (whose wife Gerard tries to seduce one night), is Ferreri isolating women in a ghetto based on his own dangerous tendency towards abstractions? Michel, the older man in Valerie's life, says to Gerard at one point, 'There are big changes happening. You have to handle women delicately these days.' Yet Michel treats his own girlfriend without much finesse.

Gerard Depardieu just before the harrowing final scene in *The Last Woman* – and Ornella Muti (above left) as Valerie

39

The Last Woman

Gerard, however, perhaps sensing that here is a man who is facing some of his own problems, but handling them rather better, is eager to listen. Later, alone with Valerie, he says that he is not clever enough to cope with the changes and adaptations necessary to become a 'whole' man. He goes out on his own, on his motorbike, wearing black leather in a way that is un-mistakeably ceremonial.

There is nowhere for Gerard to go. He ends up in the compound of the refinery where he works, a tiny figure dwarfed by high blocks. Back at the flat, Gerard wakes Petey and cuddles him in the kitchen. Valerie comes in and takes the toddler from him. There is a slight switch-off in Gerard's eyes as the girl and *his* child walk out of the room. He swallows half a bottle of wine and picks up the meat-carver.

Ferreri has explained Gerard's self-castration in the following way: 'He wants to take on a different physiognomy: he wants to be a man and not a phallus.' The ending to his film is as bleak as anything in contemporary Italian cinema, a cinema much versed in the delineation of desperation. The camera frames Gerard, bloodied and sobbing, and the girl holding the child. He also sobs, as though sensing the anguish of his mutilated father. In this shot, Ferreri encapsulates his own rage at the failure of modern society to contain the tribe's cohesion. But the very biblical quality of this rage somehow carries any solution to social frustration beyond his reach. The weakness of *The Last Woman* is its failure to locate the potential sources of harmony between men and women and the interludes of humour and tenderness. This makes the film, for all its visual power and Depardieu's excellent interpretation of the character – half braggart half little wounded soldier – a little too schematic.

Ferreri's cinema, more than any other Italian film-maker's of the present period, has all the limited yet vivid fascination of a Tarot pack. Images recur like symbols from the major arcana, reshuffled in different series but always essentially repetitive, and perhaps unsatisfactory because they are ambiguous. He is, in fact, the most Crowleyan of Italian *cinéastes*, seeking a refuge behind a barricade of often arresting signs which, after the shock of impact fades, remain opaque. He stands at the opposite pole of the expressive spectrum to, say, Francesco Rosi, who struggles always to make his meaning absolutely clear, on the democratic level of participating ideas, perceptions and denunciations. For all its immediate urgency, Ferreri's cinema seems too frequently marred by egocentric inaccessibility. What Ferreri ignores, *vis à vis* his own films, is their inherent lack of appeal for the hypothetical mass audience which he feels is being deprived of them. Yet his films are too nihilistic to appeal to any but the self-consciously alienated. The mass of Italians would probably prefer an Alberto Sordi comedy directed by Monicelli or Comencini.

Ferreri's difficulty, in fact, has been that he has somehow always fallen between two stools. On the one hand, he lacks the realism and humour of Italy's most appreciated comedy directors, Monicelli, Comencini and Risi. On the other hand, his own artistic dignity is not equal to that of Italy's foremost 'personal' film-makers, such as Rosi, Fellini or Antonioni (both Rosi and Fellini, incidentally, are capable of filling cinemas in the major cities with a new film). There is, then, in Ferreri a certain sense of spoiled expectations, what one ex-anarchist interviewed in a recent *New Society* article (23rd November 1979) termed 'the politics of individual paranoia'.

Critics have often praised his cinema but in the wrong terms. *Vide* Alberto Cattini (*Cinéaste* Vol VI No 2) – 'Ferreri has always been hindered by producers and distributors because his world-view is so deeply and lucidly anarchic.' Or Gino Buscaglia (*Framework* No 2) – 'The system is skinned to the bone and analysed under the microscope of the grotesque' (in his cinema). To conclude this assessment of Ferreri, however, one might cite the comment of another critic, from another era: 'In his hands the ease and fluency of the Baroque are gone; and the results only come from an intensity of pain ... in a private language or cipher as full of symbols, as fiery of feeling as El Greco.' This is Sacheverell Sitwell describing the effect of Borromini's paintings in his book on Southern Baroque. The 'pain' and the privacy of language also seem to apply to Ferreri.

MARCO BELLOCCHIO
THE AUTHOR

2. Distance and Didactics: Marco Bellochio and Liliana Cavani

One by one, the institutions have been inspected in Bellochio's cinema, and indicted: the family, in *Fists in the Pocket* (1966); party politics, in *China is Near* (1967); the press, in *Slap the Monster on Page One* (1972); the Church, in *In the Name of the Father* (1972); and the Army, in *Victory March* (1976). In addition to the feature films, Bellochio has worked as part of a film-making cooperative, known as the *Centofiori*, in collaboration with Silvano Agosti and the critics Sandro Petraglia and Stefano Rulli. The coop has produced two significant documentaries to date, a study of Italian mental hospitals made in 1975, *Nessuno e Tutto* (later released in a shortened form under the title *Matti di Slegare/Madmen to Release*), and *La Macchina Cinema* (*The Cinema Machine*), made in 1978, an inquiry into the corrosive influence of commercial cinema and what can be done about it.

Bellochio has also, at intervals, involved himself in theatre production. In 1969 he produced *Timon of Athens* at Milan's Piccolo Theatre and in 1977 he made a tele-film of Chekhov's *The Seagull* for RAI's second channel. In Sandro Bernardi's useful monograph on Bellochio (*La Nuova Italia* series, 1977) Bellochio comments, 'My work has always oscillated between cinema, theatre and documentary in order to avoid being constricted in the role of director.' From some film-makers, this might sound pretentious; from Bellochio, however, whose output has been relatively sparse but always rigorously conceived and executed, the remark is wholly valid.

Bellochio's great strength is that he never leaves the world behind him. There is no trace in his cinema of that self-congratulatory intellectualism which emerges from Bertolucci's movies, say, and indeed the parallel with Bertolucci is useful, up to a point, for they were born within two years of one another (Bellochio, born in 1939, is the older), both come from Northern Italy, and both made their initial impact on Italian cinema at much the same time. When *Before the Revolution* and *Fists in the Pocket* burst on the European festival circuit in 1965/6 there did seem a case for arguing a new dynamism and vitality in Italian cinema akin to that of the early neo-realist films of De Sica and Rossellini.

Bellochio studied film at the Centro Sperimentale (like Liliana Cavani) and then at the Slade in London. Before making his debut feature he had completed a couple of short films, one a spoof-gangster movie, the other a documentary about a graveyard. *Fists in the Pocket* was realized on a slender budget in the mountainous region of Bobbio, in and around Bellochio's home area, and shows a state of rapidly deteriorating family relations. Mother is old and blind, and removed from daily realities and pressures. Augusto, the oldest of the three sons, is reluctant head of the family. He is planning to get married before long and move into town. His two younger brothers are Sandro (Ale), who is given to occasional epileptic bouts but seems very alert, if somewhat unnerving in the unpredictability of his actions and responses to familial situations, and then the mentally retarded, youngest brother, Leone. The family is completed by a daughter, of roughly Ale's age, named Giuliana. She seems over-attached to both her older brothers. This, then, is Italian bourgeois family life as Bellochio, at the age of 27, sees it.

Jean Delmas, writing in *Jeune Cinéma* (No 27/8, February 1968), described Ale as a 'diabolus ex machina', a phrase which can hardly be bettered. Having committed matricide and then fratricide, having celebrated the lifting of these ancient family fetters by making love to his sister (who becomes briefly catatonic), Ale suffers a fatal attack of epilepsy while listening to Verdi's *Sempre libero*. The film ends on Ale's no longer twitching, frozen body.

Like certain other Bellochio heroes, notably Angelo in *In the Name of the Father* and Captain Asciutto in *Victory March*, Ale's hypermania, which might superficially be mistaken for a form of exhilaration, is a side effect of his desperation. His tragedy is that he is a little too intelligent not to see his older brother's purely materialistic ambitions as paltry, but not stable enough to channel his intelligence in an alternative direction. That Bellochio sympathizes with Ale's all too real problem (if not with Ale's all too melodramatic solution) is shown in a telling little scene depicting a card game which takes place between Ale and Augusto, with Giuliana watching. Augusto cheats while Ale is away from the table, and Giuliana tells Ale. 'I know,' he says. At this moment Ale seems mysteriously strengthened (he has already killed his mother) and Augusto is revealed in a darker light. His easy materialism is a form of opportunism.

Bellochio's second feature film, *China is Near*, while its iconoclasm is just as evident, is a good deal more witty. The *grand guignol* elements are entirely absent; the social targets are precisely focused. It is a film which shows how sexual and political opportunism intermingle, as will be clear from a brief synopsis:

As in *Fists in the Pocket*, the main protagonists are siblings. The Malvezzi family comprises Vittorio (35), Elena (30) and Camillo (17), who live together in the family mansion in Rome, protected from exigency by wealth and lineage. Vittorio, the present Count, is running in the city council elections as a Socialist candidate ('Socialist' in Italian politics implies social democratic liberalism). Vittorio interprets his new-found populism on an erotic as well as a careerist level: he has just taken his secretary, Giovanna, as a lover, not knowing that her boyfriend is Carlo, Vittorio's political agent. Carlo's opportunism is more naked than his employer's, it soon becomes evident. 'The more I have to put up with the more I want,' he tells Giovanna; it might be the watchword of all the characters. When Carlo arrives at the Malvezzi house, he makes an immediate play for Elena, who is tired of her current bedmate and quite responsive. Carlo's motives for seducing the mistress of the household are primarily but not wholly pragmatic. 'It's not that I like her money and not her,' he confesses, '... I like them both together.'

The twists to this potentially farcical situation – two pairs of lovers getting their various wires crossed – intensify when Elena gets pregnant and wants to get rid of it. Giovanna, on the other hand, isn't pregnant and wants to be, in order to entrap Vittorio into an advantageous marriage. The attitudes of both women are thus shown to be central to the story's development and resolution. This is unusual in a Bellochio film. Not until *Victory March* does the role of a woman become central in the contemporary sense of self-determining. The prominent role of the women in *China is Near*, aligned with the film's consistent wit and lucidity, suggests that Bellochio's co-writer and 'artistic collaborator', Elda Tattoli (who also plays Elena and later became a director herself), is important to the overall effectiveness of the project.

There are three well detailed political scenes in *China is Near*, each with considerable intelligence behind it. The first occurs when Vittorio, accompanied by Carlo, tries to make a campaign speech at a small town, only to find no audience has been mustered by the local agent, a farmer who that day is sowing. Carlo points out that the hustings involves such minor setbacks. Vittorio snaps, 'I'm no kamikaze for the Socialist Party and I don't want to make myself ridiculous.' He then makes a basic error, chasing a small boy who has grabbed his speech notes and shaking him violently. This causes a riot in which Vittorio is beaten up and has his new Mercedes damaged. Later, we see his style of haranguing those who are indifferent to his political ambitions. At home with two old maiden aunts, whose votes he needs, he positively bullies them into supporting him, capping the tirade by telling the maid, 'If you don't vote for me, I'll fire you.'

Vittorio's political duplicity is suggested most conclusively, however, in a confrontation which occurs between him and an elderly nightwatchman who surprises Camillo, Vittorio's young, Maoist brother, daubing 'China is Near' over the walls of the Socialist Party headquarters. The old man sticks to his guns, insisting that Camillo clean off the Maoisms. Vittorio bullies him into taking money to forget it, then threatens him with accepting a bribe. It is social sadism of the sort we see so prominently in the behaviour of the police chief in Petrie's *Investigation of a Citizen Above Suspicion*.

In the penultimate scene, in which Vittorio finally makes his 'rousing' speech before a packed audience at the Philodramatic Theatre, further sabotage devised by Camillo is introduced, in the form of two large dogs and a cat, which chase Vittorio offstage, to laughter from his prospective voters. Political ambitions of this puny, self-promoting sort are just one strand in a universal corruption of farcical dimensions. Thirteen years after it was made, *China is Near* remains impressive for its verbal and visual humour and great clarity of psychological observation. Tommaso Chiaretti cites Bunuel as an influence on Bellochio (a comparison Bellochio does not reject – who would?) but the lightness of touch and unimpeachable levelheadedness of *China is Near* is even reminiscent of Oscar Wilde. 'I imposed on myself very precise limits, playing a certain game,' Bellochio commented (*Jeune Cinéma* 27/8, February 1968). 'It is a very rational film, with many coincidences combined in an unrealistic, calculated manner.'

The first of Bellochio's two feature films made in 1972 was *Sbatti Il Monstro in Prima Pagina* and came about in an unpremeditated manner. The script had been written by Goffredo Fofi and Sergio Donati for the latter to direct. Illness prevented Donati from realizing the project, and Bellochio was called in and given only two weeks to revise the screenplay and make production preparations. *Monster*, which shows signs of the haste with which Bellochio had to assemble his production, has never been made available to a UK audience and can, thus, in a sense, be regarded as Bellochio's 'missing' movie.

In its theme, *Monster* bears certain affinities to Vittorio de Seta's *Un Uomo a Meta* (*A Man in Half*) made in 1966, a surprising shift of direction from De Seta's best-known film, *The Bandits of Orgosolo*. Both films examine the pressures brought to bear upon individual journalists by the vested interests which control the Press. In stylistic terms, however, the two films part company. De Seta's study is slow-moving and introspective, depicting the inner crisis of his journalist-hero. Bellochio's film is faster-paced, makes use (for the first time in a feature of his) of documentary shots as inserts, and in its intellectual stance is more directly denunciatory of the repressive power of the Italian press.

The story concerns a newspaper editor (played by Gian Maria Volonte) who is obliged by his proprietor to run a sex and violence story in a prominent position in the paper. The scandal concerns the rape and murder of an eminent physician's daughter. The killing has to be slanted to suggest that Left-wing terrorists are behind it. This way, so it is argued by the proprietor and his political allies, distraction will be offered to the sensation-hungry readers, who will fail to notice the paper's political tactics – harassing Left groups during the run-up to the elections. Unfortunately, the journalist who investigates the case, like Inspector Rogas in Rosi's *Illustrious Corpses*, comes up with an entirely different story from the one expected of him by the powers that employ him. He discovers that the girl was murdered by a school janitor. The editor has to fire him. The elections are more important than factual exposés.

If *Slap the Monster on Page One* is Bellochio's most turbulent movie, especially in its use of documentary inserts showing Almirante, former fascist chief of Saló, now involved in the neo-fascist movement, one should consider the fact that Italy, in 1972, was rife with urban violence, along with its usual concomitant, paranoia. The Feltrinelli case, for example, pointed a clear parallel in life with Bellochio's thinly fictionalized denunciation of Press and police corruption. Feltrinelli was a well-known Milan publisher with emphatic Leftist sympathies whose corpse was found in the outskirts of Milan beneath a pylon. This death, too, occurred on the eve of elections. Feltrinelli was posthumously smeared by the Press, who alleged he was a terrorist leader.

Bellochio's most overtly political feature film, for all its turbulence, stops short of stridency. As always, Bellochio is concerned with keeping a certain distance between political events and the characters of his drama. The editor, for instance, while presented as a man in the pay of negative forces, is not personally evil. He is shown to be intelligent, and good at his job, professionally if not ethically. We are in a curious position *vis à vis* this vital character. We respect him for his personal qualities while wishing to dissociate ourselves from his invidious professional position. By provoking this sense in us of an uneasy quandary, Bellochio maintains his critical stance towards not merely the latent authoritarianism of Italian institutions but also the hazard of degrading cinema into a propagandist medium.

In the Name of the Father, Bellochio's film about a Catholic seminary in the late 1950s, contains many odd, somewhat puerile subversions. Working with Tattoli five years earlier, Bellochio had contained his portraits of anti-Catholic adolescents in the sub-plot involving the younger Malvezzi sibling, Camillo, who was also in a seminary and much given to abusing the priests, as these pupils are. Now the theme of adolescent anti-clericalism is promoted to a main position, with a student hero, Angelo (Yves Beneyton), who has that dangerous relish for violent action shown by Malcolm McDowell in Lindsay Anderson's *If*, about a British boarding school. Angelo spearheads the pupils' conflicts with the priests and most of the critical action crystallizes around his presence.

Yet there is another strand to the film, concerning the manual workers at the institution, a passively resigned bunch whose militancy, up until now, has not amounted to more than spitting in the soup tureens back in the kitchens. Their sudden stirring into collective action is the film's directly political aspect, though it should be said that part of the weakness in this element derives from Lou Castel's surprisingly sluggish performance as Salvatore, the self-elected spokesman of the menials, who leads them in their bid for better pay and working conditions.

At the heart of the power struggle contained in the movie, however, is the conflict between Angelo and the director of the institution, a bland priest who has seen it all before. When the boys go too far, he tells them, wearily, 'All you're waiting for is to be out. But your characters are formed here, even if you don't realize it.'

This familiar Jesuit assertion is important to Bellochio, one senses. His films bear out the weight of the Jesuit dogma in one way or another. Bellochio's heroes are unbalanced in ways which suggest the constraints of an antipathetic institution in the background, be it family, Church, Army or

Yves Beneyton in *In the Name of the Father*

school. There is about men like Ale, Angelo and Asciutto an hysterical tendency which signals the past prohibition of warmth, tenderness and creativity. How valid is the priest's sombre warning, applied to these boys? More than a little valid. These pupils are a brawling bunch, already rehearsing the various forms of oppression among themselves that they will shortly be free to inflict on those in their employment or jurisdiction. For the world they are soon to enter is secular, materialist and technological, and will offer them unbridled opportunities to practise the devious multiple tyrannies known as 'healthy competition'.

Even Angelo, the most sharply individualized of the boys, is shown to be little more than a petty tyrant, despite his extra vitality. 'His aspiration to power, a temptation intrinsic to western intellectuals, is also mine,' Bellochio has commented (*Sight and Sound*, Autumn 1973). He sees Angelo's most negative facet as 'his total contempt for people, combined with the side of him which sees progress through authority'. Angelo's fanaticism – his crazy power games, his blasphemies, his acts of vandalism – represent a more socially aware development of Ale's tribal terrorism in *Fists in the Pocket*.

'Some critics have seen fascist tendencies in *Father*,' Bellochio remarked, 'perhaps because Angelo's charm goes beyond the limits.' Perhaps there are authoritarian fantasies at work in the movie, a will to power more nakedly divulged than Ale's precisely because it is less psychopathic. Angelo, the malign subversive type, crops up again in the portrait of Sciabola in *Salto nel Vuoto*.

Power in the seminary works in a crude, archaic manner. The right to bargain does not exist for the workers. They too must 'obey', but not in order to become leaders, simply to remain subservient. A further feature of the unalloyed power preoccupation arises from the fact that here there are no women. In the absence of real women, as other directors in Italian cinema besides Bellochio have shown (Wertmuller being a notable example) men tend to fall back on madonna-worship. A madonna appears before several boys at their most heated moments. Even in her wimple, with her demurely downcast orbs, she retains a striking resemblance to Anna Karina!

Anna Karina in *In the Name of the Father*

There is only one scene in the movie depicting women as they are. This occurs outside the institution, at the home of Franco, another pupil. Angelo and Franco are rehearsing their school show in the sitting room behind locked doors. Franco's mother (Laura Betti) fumes outside, convinced that they are sodomizing one another, like the young 'beasts' they really are. This sardonic vignette effectively introduces a touch of Italian 'mommaism' to balance the mystical madonna motif.

Stylistically, *In the Name of the Father* remains Bellochio's most elaborately devised production. In his use of low key lighting, black uniforms, and strange patterns of stained glass (later borrowed by Argento in his meretricious movie of seminarian *grand guignol*, *Suspiria*) Bellochio creates a lowering, baroque atmosphere very different from the monochromatic severity of his earlier movies. It is more than possible, too, that *In the Name of the Father* represents for Bellochio a watershed in terms other than style alone, a film in which he came to terms more thoroughly than hitherto with his own anger. His curious letters to Pasolini suggest a releasing process of this sort. 'The value of anger,' he told Pasolini, is that 'recovery is made through it,' an unexceptional point which merely attests to the Greek idea of catharsis. His most arresting comment in this intriguing correspondence, however, occurs when he says, 'I don't believe in scandalizing. Scandal is a sentimental subversion . . . the work of the satirist consists of attacking to free others from the need to attack.'

Here we see Bellochio's great strength as an *auteur* and also, perhaps, his minor weakness. What could be more pertinent than to tell Pasolini, that cardinal of cinematic scandal, that he is compelled by a 'sentimental subversion'? Yet to suggest that the satirist in *his* attacks frees others from the need to do the same is surely less penetrating? This suggests that tinge of missionary complacency which occasionally mars Bellochio's satire. The satirist attacks in order to liberate others so that they may find the poise to attack; he can never do their resisting for them.

Victory March again shows us sadistic pranking as a major element in this study of modern Army life in Italy, but now counter-poised with psychological portraits of considerable delicacy. *Victory March* suggests a new maturity of outlook, in which Bellochio gets closer to his three main characters, the young recruit, Passeri (Patrick Dewaere), the Army Captain, Asciutto (Franco Nero), and Asciutto's reluctant wife (Miou Miou).

Tenderness and anger coexist in the film's central relationship, which is the potentially troilist situation arising from the inter-relationships of two men and a woman in and around a provincial Army camp. Passeri is a new recruit, perhaps an only child, somewhat precious and mother-orientated, as is made plain in an early scene when he is out on the town and instead of joining the other 'rookies' at the brothel slips off to make a desperate long-distance call of distress to his mother. It does not look as though Passeri and Army regimen are going to mix. Initially we see him brutalized on the parade ground and also in the barracks by older conscripts, known as *nonni* (literally,

Lou Castel (right) in *In the Name of the Father*

'grandads'). What happens to Passeri, though, is that he gets 'spotted' by Asciutto as potential officer material. Asciutto proceeds to toughen him up a bit. Asciutto then shows his trust in his protégé by entrusting him with a delicate personal mission.

The Captain's wife, Rosanna, is making him suspicious. He wants to know where she goes in town and orders Passeri to follow her. Passeri does so, reluctantly, and rapidly discovers that Rosanna indulges in shop-lifting and also visits a lover, a junior officer named Baio who is almost as bombastic as her husband, but has the saving virtue of being potent.

One day, during the surveillance procedure which has by now become almost routine, Passeri follows Rosanna to a cinema and sees her weeping. Outside, he comforts her. Rosanna is touched and, only half-jokingly, suggests that they should run away together. They drive out to the country. Passeri tries to convince Rosanna that her husband loves her, albeit in his contorted way. But Rosanna knows better. Contact with Passeri helps her decide: she is ready to leave Asciutto, but first she must extricate herself from the affair with Baio.

This she does, quite abruptly, but Baio retaliates by phoning Asciutto

Miou-Miou and Franco Nero in *Victory March*

anonymously to tip him off about Rosanna's infidelity. Asciutto is so furious that he waits up for Rosanna, and threatens her with a pistol. Rosanna walks out for good and Asciutto implores Passeri to get her back for him. Passeri tries to intercede with Rosanna but she is adamant about leaving. Now seriously unhinged, Asciutto plans a raid on the camp sentry-box.

That night Passeri watches as Asciutto conducts his mad raid on the sentry-post. The guard shoots him dead when he advances, ignoring the challenge. He then fires a second shot in the air. As Passeri kneels by Asciutto's body, the guard approaches to tell him, 'You saw me fire a warning shot first?' Passeri agrees to corroborate this distortion. With his protector dead, he now has to survive in his peer group, or risk total ostracism.

This, in bald outline, is the shape of the narrative. *Victory March*, however, has a number of felicities over and above the strength of its storyline. For a start, the humour, which is particularly pointed in the scenes between Rosanna and Baio, whose 'love-nest' is his room, filled with erotic magazines and movies. Baio tape records the sound of Rosanna's orgasm and, after she has walked out on him, fearful that his fellow officers will ridicule him for losing a woman, he plays back the tape as they lurk below his bedroom window. Unfortunately, the other men happen to know he's on his own, which makes his disgrace doubly ridiculous.

Bellochio also examines the dangers of *macho* attitudes and habits within the marriage. Rosanna is demoralized by Asciutto's insensitive domination. Equally inappropriate is Passeri's blind loyalty to his officer, regardless of the human situation – misery – which Asciutto provokes. Just how violent *both* men can be Bellochio sketches vividly in the scene outside the barracks, in which Asciutto and Passeri set upon a street homosexual. Egged on by the Captain, Passeri goes berserk, giving the deviant a severe beating before he is called off.

The connection between Fascism and sexual abnormality is all too clear in the two men, and also in the frequent barracks sadism, as when the *nonni* saturate Passeri's bed with plastic bags filled with noxious fluids. Bellochio evidently intends to show, like Pasolini in *Salò*, that sadism is the face of advanced sexual insecurity. Thus, when Asciutto tells Passeri that he would like a son because it would be 'Someone in your hands – you could train him from birth,' he reveals the extent to which he is given over to the authoritarian codes which suffuse institutional life, especially perhaps the Armed Services.

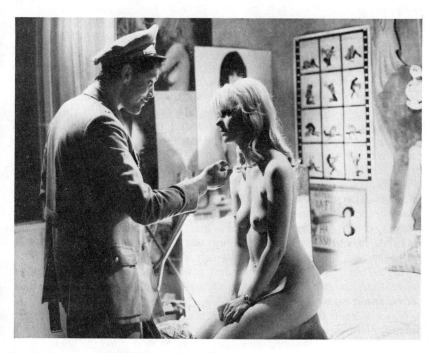

Patrick Dewaere and Miou-Miou in *Victory March*

At the outset, Passeri has complained to Guancia that 'the collective life is killing me ... never a moment to be alone'. At the end, he returns to the group life, after his curious interval as middleman in the tormented marriage of his sad Captain. Guancia has told him, 'Try to be less egotistical. You think you're the only one who suffers.' The group life, as displayed in this portrayal of the petty atrocities of the barracks, reveals the ferocious process by which covert homosexuality is converted into humiliation rituals.

Is it plausible that conscript life in a modern European army is as black as Bellochio and co-writer Sergio Barzini paint it? Are the indignities we see, which seem to derive from the world of de Sade or Salò, really in evidence today? From a report in the *Guardian* (21st September 1979) comes an account of an ordeal undergone by a British army recruit who was 'tried' by a junior officer, Lieutenant James Deedes, blindfolded and 'executed'. Deedes was cashiered for his offence. It could be argued that he got off lightly. Nothing changes, it seems, in the Army.

The distinct virtue of *Victory March* is that it leaves no unsatisfactory moral ambiguities, as did *Fists in the Pocket* and *In the Name of the Father*. Again, Bellochio has scrutinized the excesses of the authoritarian personality, but now it is clearer what he thinks about it. But, oddly, Bellochio, now revealing a new crispness and open-heartedness towards human problems, turned away from the contemporary milieu in his next project, *The Seagull*.

Chekhov's play was termed a comedy by the author. Bellochio's version is a more sombre affair, closer to Visconti's late cinema, *Death in Venice*, say, or *Conversation Piece*. It is a play hinging upon several sexual conflicts and two prophetic warnings. The nub of the drama is the mother/son conflict beloved of other Italian *cinéastes*, notably Pasolini and Bertolucci. Konstantin (Remo Girone) is a young poet who writes verse dramas and is in love with his mother, Irina (Laura Betti). Irina, a famous actress with a middle-aged novelist/boyfriend, Trigorin, has no more time to devote to Konstantin's tormented introspections than Caterina had for Joey's in *La Luna*. When Konstantin puts on one of his shambling verse dramas in the garden of their country estate, using a shy girl from the neighbouring estate, Nina, as his solo performer, Irina, in the audience, refuses to become mournful, as is expected. This is the old uncaring mother mechanism which Ale first exposed in *Fists in the Pocket* – 'Mummy, I'm unhappy.' 'Do you want a sweet, dear?' Ale responded to this maternal cruelty by killing her, like Oedipus. In Chekhov, however, things are more plausibly distanced – we see the mother destroying the son, which is surely more common.

Just as unrequited mother-love destroys Konstantin so father-daughter infatuation almost destroys Nina, whom Trigorin mistreats during their period living together in Moscow.

Why is Bellochi attracted to the play's darker elements? Bellochio himself, discussing the movie (*Jeune Cinéma*, No 118), remarked on the sense of pessimism he experienced during shooting. 'The text led me backward. I experienced it as a kind of regression ... I had to discover my adolescence ...

everything took place under the sign of madness, of jealousy and impulsive anger ... A kind of detachment took hold of me, as if I wanted to keep the images as far away as possible.'

Since *The Seagull*, Bellochio has completed two more films: in 1978, a documentary inquiry produced by the same film cooperative which earlier made *Matti di Slegare*, *La Macchina Cinema* (*The Cinema Machine*), which remains unscreened in the UK at the present time, and a feature film, finished in Rome in the summer of 1979, *Salto nel Vuoto* (*Leap into the Void*), with Michel Piccoli and Anouk Aimée in the leading roles.

First, though, *The Cinema Machine*, in which Bellochio's openness of perspective is also evident. The documentary is ambitiously conceived by Bellochio and his three colleagues in the *Centofiori* cooperative (Silvano Agosti, Sandro Petraglia and Stefano Rulli). The press release of the cooperative announces, 'The cinema is a machine that produces goods ... we have not provided a forum for ... the "masters" of the cinema, but rather for those who have found themselves excluded by the machine, those who do not accept its logic and its blackmail ... our starting points were: our own unemployed friends, a certain personal unease felt while attending the ceremonies of the film world, but also the fascination that many – ourselves included – still find in film-making.' This statement contains two definitions of cinema – the commercial industry, and also cinema as a state of awareness and hope, 'the fascination' which afflicts or enlivens those who constitute the cinema-world of this study.

The documentary begins and ends by showing aspects of the cinema industry itself. The first shots show Marco Ferreri at work in a studio at Cine Citta rehearsing scenes from *Ciao Maschler*, with his 'international' stars Gerard Depardieu and Gail Lawrence. We quickly cut to Ceccano, in the country, where an amateur film-maker, ex-boxer Tony De Bonis, is holding a prize-giving ceremony in the town hall for his last 8 mm. movie, *The Partisan*. What is curious about this scene is that it suggests on De Bonis's part a desire to imitate the razzmatazz of commercial film-making rather than reject it. The film is shown, and depicts De Bonis and friends hamming it up exuberantly as brave partisans fighting evil *fascisti* in their back yards.

We then meet a rural teacher who wants to make nature documentaries about the night-life in his local woodland. He turns out to be no less interested in the grand Freudian themes than professional *maestri* like Pasolini and Bertolucci, attributing a period of impotence in his life to his late father's frequent infidelities when he was a child. The teacher talks of a film he has made about his own momentous sexual experiences – 'that most beautiful thing', as he terms it. The scenes we see are Olmi-like shots of a peasants' open-air dance, followed by a young couple making love by a lake. 'I have to thank Christian culture,' the film-maker observes, attributing his motivation for filming to 'the sense of sin, which was a powerful propellor'. Once again, the emphasis on a non-political motivation for making films sets up the parallel in the professional cinema with Olmi.

Cut to Giacomo, an elderly, somewhat embittered man who once tried and failed to run a film club. It foundered, he asserts, because of his wife – 'She thought I was mad. She begrudged the money.'

Next we meet a young actor who played a role in *In the Name of the Father* seven years before. He has returned to the provinces, disillusioned with the internecine film world of Rome, which he describes as 'mental death'. If De Bonis, with his prize-giving awards, reflects the great fascination of many amateur film-makers with the 'glitter' of professional cinema, this young actor is much closer to reflecting the secret dreams of those who have worked on this documentary. For no one who has direct experience of the cinema machine can fail to harbour on occasion the secret dream of turning one's back on the machine and returning to a personal simplicity. We see the actor's big scene in *In the Name of the Father* (playing a pupil who was ordered by Angelo to hang from a bar in the gymnasium for one hour, and failed to pass the test – by two minutes). He is now selling handbags at country markets. Is he right in his decision to turn his back? The film does not answer this question. It does, however, trigger it. By turning his back on the cinema machine he gives in to his own sense of disappointment, perhaps. Is it in fact a version of hurt pride which lies behind his decision?

This important question is addressed again in *The Cinema Machine* when we see another ex-actor, now involved in community work. From a largish role in *Padre Padrone* he has returned to the streets. Now he is working with a group of unemployed people, trying to landscape stretches of wasteland in the city. Through the eyes of this man we watch his brother practising jabs and lunges in a karate class. The ex-actor explains that these karate trainees have the urge 'to belong, not to die of loneliness ... They're not organized, that's the trouble. They have to work out their violence like this. They don't use it in a more profitable way.'

Next, a couple of producers of sex films are seen trying to persuade coy girls to do their 'screentests' naked. Their methods of cajoling the girls are ingenious – 'Do you know the cinema is in a state of crisis?' demands one producer, thus establishing a marvellously spurious connection between stripping and artistic courage. The girls don't entirely buy it but they are prevailed upon to show their breasts, and sit there giggling. Two homosexuals are more readily enticed into self-exposure, however; they even do a little dance in the manner of those 'bare boys dancing' scenes so favoured by Liliana Cavani in her later movies.

The penultimate sequence revolves around a one-time star named Daniella Rocca, who made great impact in her role in Pietro Germi's *Divorce Italian-Style* in the early Sixties, being compared at the time to Loren and Cardinale. Unlike the others interviewed in *The Cinema Machine*, Daniella Rocca reached the 'pinnacle of success' and has of course fallen lower. She has been in mental hospitals, in the bankruptcy courts, and now lives in a denuded flat, with the power and telephone disconnected, surrounded by old movie mementoes.

The difference between Rocca the star and Rocca now is this: once she was locked in a gilded cage, now she is out on her own, more fragmented, totally impoverished, but open. She prefers to remain poor, she says, than to swallow her pride by working in an office. 'The cinema will come back to me. If I don't kill myself.' Bellochio tells her she has never stopped acting, she is still acting – at this moment. His tact and sensitivity is agreeable.

What is clear from *Salto nel Vuoto* (*Leap into the Void*) is that Bellochio is steadily developing as an artist. His increase in range and control, from feature to feature, is never pyrotechnical but is certain for all that. The theme of *Salto nel Vuoto* is once again the claustrophobic power of family relationships. In this sense the film marks a return to the features of the Sixties, which dealt primarily with conflicts *inside* the family. *Salto nel Vuoto*, then, resumes the old themes which interest him – the fact that power ambitions tend to trivialize the personality, and that women in male chauvinist households are almost inevitably oppressed, but unlike *Fists in the Pocket* and *China is Near*, *Salto nel Vuoto* is neither operatic nor satiric.

What Bellochio achieves, in essential terms, is an awareness of certain pertinent social and political shifts. He shows us not only Judge Ponticelli (Michel Piccoli) locked in his odd love-hate relationship with his sister, Marta (Anouk Aimée), but also Ponticelli's adversary, the fringe theatre activist

Anouk Aimée in *A Leap in the Void*

named Sciabola (Michel Placido), who operates from a raft-theatre in the Tiber, and fascinates the Judge with his nonchalant anarchism. 'Money doesn't matter as long as it circulates,' he tells the Judge at one juncture, in between practice bouts of fire-eating – feats which intimidate Ponticelli and fascinate his sister.

The Judge first meets Sciabola when he is called to investigate the circumstances of a suicide. The victim, an ex-girlfriend of Sciabola's, has jumped from a high window. At the beginning of the film we see the Judge leaning over the parapet from which she hurled herself. At the end of the film, when the Judge kills himself in like manner, we understand the personal darkness which underlay his morbid fascination.

The Judge questions Sciabola, whose telephone conversation with the dead girl has been tape recorded. He suggests that Sciabola's verbal abuse helped to trigger the girl's suicide. Yet behind his professional interest in Sciabola lies the kind of awe so often experienced by the apparently respectable for the apparently shamanistic. When his sister Marta departs abruptly from a family wedding, tired of being cast yet again as maiden aunt in residence, Ponticelli follows her down the Lungotevere. He shows her the theatre-raft, and introduces her to Sciabola, leaving her with the actor in the midst of rehearsals. What develops between Marta and Sciabola is never shown explicitly yet there are later hints that Marta has become the actor's mistress. In one telling scene we see the Judge return home to find that Marta has been ironing (normally Anna the maid's function in this household). The shirts are not his, but Sciabola's.

The complex ambivalence of Ponticelli's feelings towards Sciabola are delineated exactly by Bellochio in the screenplay which he co-authored with Piero Natoli and Vincenzo Gerami. Ponticelli has virtually offered his sister to Sciabola. Her behaviour has been strange – the Judge describes it as 'menopausal'. She skulks in her room, talking to herself and giggling hysterically – that same hysteria which was so much a feature of the much younger siblings in *Fists in the Pocket*. Ponticelli and Marta are indeed akin to Ale and Giuliana of that film, but grown twenty years older. Ponticelli offers Marta to the actor out of desperation, for his morbidity has provoked Marta's hysterias, and he senses it. 'Have the courage to dream,' Sciabola once tells him. 'If my dreams came true the cemeteries would be overflowing,' retorts the Judge savagely. Towards the end of the film, when Marta has found enough courage to agree to spend a weekend away from him, with Anna at the seaside, Ponticelli comments disgustedly on the notion. 'It's so crowded, so dangerous. The water is polluted.' The fear in which he holds life has finally been challenged by Marta. When he gives her money to spend while away, he has to point out to her that the fifty thousand lire note resembles one of lower denomination. An acute detail, this, suggesting not only Marta's lack of experience at handling money but also the Judge's hope that she will prove helpless.

For evidence of Bellochio's mastery of control over *mise en scène* as well as screenplay we might cite the penultimate scene, in which Sciabola and his theatrical colleagues vandalize the Judge's apartment, swilling his collection of miniature cognacs and finally urinating on his desk. These images of wanton destruction are accompanied by others, for in the background several old-fashioned children are hovering, watching delightedly. When the actors depart they step forward to try on the items of costume left behind. Then they too vanish, as Marta enters the apartment. The use of a sequence-plan shot, in which each exit and entrance is unified within one composition, is intelligent not only visually but also enriches Bellochio's statement here. Those children are of course the Ponticelli siblings in childhood. There have been earlier flashbacks to scenes from the Judge's youth, including harrowing shots of an older brother, apparently handicapped, perhaps psychotic, like Ale in *Fists in the Pocket*. To compare Bellochio's *mise en scène* in this scene with Brusati's more pedestrian use of childhood revocations in *Non Dimenticare Venezia* (see the final chapter) is to realize his mastery.

Still more startling visually is Bellochio's handling of the final scene in the movie, in which the Judge scurries about the new empty apartment. For the first time we perceive its peculiar topography – it appears to exist as a circular structure around the central hallway. Wherever he goes, the merciless camera follows, catching him cringing in bare corners, and flushing him on to the next momentary refuge. There seems no escape. The Judge mutters, 'I can do what I like in my own house,' and, almost as an afterthought, crabs to a window and is gone, over the balcony.

Immediately we cut to Marta, in bed at Anna's house. Through the open window the sound of the sea is murmurous. Marta wakes with a jolt, as though she has experienced her brother's mortal impact on the city pavement. She climbs out of her bed and climbs in next to Anna's small son, sleeping opposite. This beautiful child (played by Piergiorgio Bellochio) has been seen in earlier scenes within the Ponticelli apartment, annoying the Judge with his ball games and his gales of laughter whenever his doting mother tickles him. On this lovely final image of Marta and the child curled sleeping together Bellochio ends his finest movie to date, in a career that has been distinguished from the first by a consistent integrity of purpose. Now in his early forties, Bellochio appears to be in fighting fettle. Provided that he can continue to raise the finance for his highly personal films, he must be seen, at this juncture, as one of Italy's foremost film-makers, and certainly the most lucidly scathing about her corrupt institutions.

Liliana Cavani was born in 1936 in Carpi. Like Fellini and Bertolucci, she comes from the Emilian region of Northern Italy. Her family was 'anti-fascist and secular', as she describes it. She is rare among Italian cinéastes of her generation in that she never repudiates her family background; although she has not been explicit on the subject, one has the sense that she had a relatively happy childhood. The vast majority of intellectuals of her generation, she contends (in a lengthy and very informative interview with Ciriaco Tiso, *La Nuova Italia* series, 1975), were brought up as Catholics, which explains their sense of inferiority and, perhaps, their attraction to authoritarian systems, whether Marxist or otherwise. Cavani studied Ancient History at Bologna University and originally intended to become an archaeologist, but her involvement with the film club at Bologna changed that. After a spell at the Centro Sperimentale in Rome she started work in television, directing a number of documentary compilations and inquiries for RAI-2, the cultural channel. This was between 1962 and 1965. Among these were *The History of the Third Reich* (1962), *The Stalinist Era* (1963), both of which were four part inquiries, followed by *Italian Housing* (1964) and several single programmes in 1965, *Marshal Petain, Women of the Resistance* and *Day of Peace*, this last being an inquiry into the present lives of several survivors from the wartime period, a crippled Englishman, a French priest and a former inmate of a Nazi concentration camp. Later in her career Cavani was able to point to the research she undertook in these documentaries as the initial inspiration for *The Night Porter*, the most famous of her films known to cinema audiences outside Italy.

Cavani's early career in television, while prolific, did not pass without difficulties, principal of which was the heavy editing inflicted on her inquiry into speculative housing in Italy, made a year before Rosi launched his fierce tirade against Neapolitan housing scandals in the more famous *Hands Across the City*. Perhaps because of the virtual censorship she had encountered, partly too because she was not herself a Catholic drawn to Catholic subjects, she hesitated before accepting the first feature film commission she was offered by RAI, a study of St Francis. Films about Francis of Assisi had already been made in Italy and out of it: Rossellini's film, usually known in the UK as *The Flowers of St Francis*, in 1951, Raffaele Pacini's *Tragic Night of Assisi*, in 1960, and Michael Curtiz's Hollywood spectacular *Francis of Assisi*, in the same year. More were to come; in the late Sixties Franco Zeffirelli also tackled the subject, calling his version *Brother Sun, Sister Moon*.

Cavani's film, *Francis of Assisi*, has been very little shown in the UK, which is a pity since in its grainy black and white footage there exists a portrait of a remarkable hippy rebel, much of the film's intensity deriving from Lou Castel's performance – Cavani ran into Castel in the corridors of the Centro Sperimentale and knew at once he would have to play the part.

Although there is a Rossellinian simplicity of style about Cavani's debut feature (shot on sixteen millimetre with the usual rough edges), her film takes us back to the Rossellini of the war years and immediately afterwards. It is ironic that at the time when Cavani seemed to reintroduce the spirit of *Rome Open City* into Italian cinema, Rossellini himself was making more urbane television documentaries, among them *The Rise to Power of Louis XIV*. Cavani's view of Francis, she avers, is very different from that of Rossellini in his earlier film. He was concerned with depicting the conventions of Catholic joy, while she wanted to portray Francis as the wandering hippy intent above all on a point of self-balance.

Ciriaco Tiso, in his generally useful monograph on Cavani in 1975, describes *Francis* as a film gripped by the central myth of nakedness. Francis strips off his clothes when his angry father, a wealthy wool merchant, drags him before a town tribunal on charges of 'filial insubordination'. Later, after the fraternity of brothers has gathered around Francis's cause, he advises one, Ruffino, to preach naked in the cathedral. At the end, as he dies, he instructs his brothers to strip him and lay him on the naked earth. This done, he dies happily.

Francis, in Cavani's hands, is an iconoclast. We see him kissing lepers, making a pilgrimage to Rome, where he distributes his money to beggars, and refusing all attempts by his less self-confident brothers to impose bureaucratic order on his inner coherence by devising a set of rules and regulations. 'But who will write it down?' he asks. 'Why do you wish to complicate something which is simple? It's not doctrine we need but love and courage.' The brothers disagree, as men will do when common sense is spoken.

This confrontation between the humble path and the bureaucratic urge Francis allays by going into the wilderness. In a cave he fasts, revealing his thoughts to a faithful brother, Leone. Francis dies in the little church at Porziuncola which he has spent many years making habitable. Canonized, his way of life becomes not revolutionary praxis but Christian precept.

Galileo, made two years later, confronts the same pressures in a man of genius. 'What stops us knowing the secrets of the universe?' wonders Galileo (Cyril Cusack), faced with his 'horizon of events' – the mystery of the solar system. Galileo's story is that of intellectual courage broken by the final test – risk of torture and death for blasphemy. Unlike Copernicus, his is the ultimate capitulation.

As in *Francis*, Cavani's approach to her portrayal is early Rossellinian rather than Brechtian: naturalistic settings, simplicity of *mise en scène*, considerable emphasis on the mundane aspects of the great man. We see Galileo at home, faced with the same pressures to normalize his life, expressed in domestic miniature, that he faces in the world outside. The woman he lives with is pregnant and wants to marry him. Galileo is evasive. He is working on a design for a new telescope; this preoccupies him to the extent that he also refuses to contribute towards his sister's dowry. He is also wary about his professional prospects. Another proponent of the new view of the solar system, Bruno Gordini, is being harassed by Papal investigators. Gordini is ordered to recant or face the stake. He refuses to recant: 'Your sentence will instil more fear in you than it does in me.' Gordini's death by fire is watched by a crowd; among the spectators is a child – held in the arms of a nun, a telling image.

With his new telescope Galileo is able to discover momentous universal truths: the movement of the planets round the sun; the flirting of his girlfriend in the street below him. When he tries to persuade a crowd in the street to take a look they are frightened. Galileo fetches a dog and fixes its muzzle to the eyepiece. Summoned to Rome, he is warned by a cardinal,

'You'll lose your soul by presumption.' Galileo replies, 'And you'll lose yours by sleeping on it.' His pilgrimage to Rome, like Francis' before him, only consolidates his sense of purpose. Here, like Francis, he faces the ire of the Vatican Council, three centuries later. He refuses to recant before a preliminary hearing. This time, he escapes with a warning but leaves believing he has a friend in one cardinal, Barberini.

When Barberini becomes Pope, Galileo makes plans to publish his findings. In the event however, the published work, *Dialogo*, is seized by officers of the Inquisition, Galileo has to seek sanctuary in the country and finally buries his box of writings under a tree. He puts himself in Barberini's hands, remarking wearily, 'I'm too old to run.'

In Rome, he is served with an Admonition and imprisoned. *Dialogo* is placed on the Index. 'Don't you realize that ambition is a sin?' inquires one young priest, refusing to look through the telescope. Galileo agrees to make a formal recantation. The night before his court appearance he has a dream in which he tells the tribunal, 'If anyone saw the devil here today I did – in some of your faces.' At his abjuration he reads the recantation blankly. He is suddenly much aged; and Cusack conveys the attrition of will with considerable subtlety. Of those who have been offered the chance to hear Galileo's ideas only his young cellmate seems to have grasped Galileo's essential message, that the pursuit of truth is holy work.

The weakness common to both these tele-films is the dialogue. Both, during the debating scenes between brothers and cardinals, become too expository, a fault Cavani had ironed out by the time she came to develop her historical documentary on Simone Weil, in the mid-Seventies. This mark of uncertainty at script level, however, is not so noticeable in the direction, which is restrained and unfussy in its use of camera and editing. What the two studies show principally is Cavani's capacity for careful research. This, throughout her career, has remained a strong feature. 'My films are not historical but are films of ideas,' Cavani has commented (*Framework* No 2, 1975).

Cannibals was produced in 1969 and was the first of Cavani's films to attract attention across the Atlantic. An American distributor who was impressed by the film at the New York Festival offered to buy the US rights if Cavani would change the last fifteen minutes. She refused, with her characteristic resolution. The story is freely adapted from Sophocles' play *Antigone* and stars Britt Ekland as Antigone and Pierre Clementi as Tiresias. The film also marks the start of Cavani's collaboration with the scenarist Italo Moscati, with whom she has since co-scripted *Milarepa*, *The Night Porter* and *Beyond Evil* – for her first two tele-films Cavani had worked with Tullio Pinelli, the novelist and Fellini's one-time script-writer.

The conception of *Cannibals* is very different from Cavani's more structured television productions. The basic misjudgement is to allow rhetoric to override narrative, sentimentality to override spirit. Ciriaco Tiso, whose evaluation of the film is warmer, comments in his monograph that he finds

Pierre Clementi and Britt Ekland in *The Cannibals*

the silences more satisfying than the polemical scenes. To this observer, the ritual passages within the film – Antigone and Tiresias burying the bodies of young rebels in a tranquil spot far from the city, or wandering naked, pursued by soldiers through a very primal garden – seem as cloying as the rhetorical outbursts seem callow.

Cavani retells the Antigone myth in a slightly futuristic setting. Milan is her location, a city Cavani dislikes for its coldness and inhospitality (where the workers have already become '*piccolo borghese*', thereby sapping the potential vitality of the city, as she sees it). Antigone's brother has been murdered, along with a number of other young rebels, in a palace uprising against King Creon. The King has ordered that the bodies be left on the streets as a warning to other potential dissidents. The penalty for touching them is death. The atmosphere is thus apocalyptic and authoritarian, like Ferreri's *Seed of Man* and Petrie's future shock extravaganza of 1966, *The Tenth Victim*.

Antigone defies the King and steals her brother's body, with the help of Tiresias, a highly conceptualized figure who arrives from the sea, eats only fish and speaks in unintelligible murmurs. They ferry the body along a river and bury it in a glade on the bank. Bread is ceremonially broken into three portions, one for the corpse, a detail suggesting pagan simplicity, perhaps, or feyness, depending on one's constitution. The drama thereafter becomes predictable. More ritual burials. Pursuit by the authorities. A symbolic sequence with the boy and girl wandering naked in The Garden. Pursuit is followed by arrest, torture in Antigone's case, and summary execution in a square, before a crowd. 'Order has been reestablished,' announces Creon. But already more bodies are being stolen, more boats are stealing up the river.

The influence of Julian Beck's Living Theatre is strongly evident in *Cannibals*, as it was in *Pigsty* and *Partner*, notably in the scenes depicting Antigone's incarceration among the patients of a mental hospital. Similar excesses mar the scenes showing Emone turning into a dissident when he sees the cruel treatment Antigone, his former fiancée, is receiving in prison. He suffers a rapid (all too schematic) transformation and ends up curled foetally in a cell screaming at his father, 'I want to become an animal, an anarchist, eccentric, rebel, delinquent, atheist, homosexual.' The litany is essential Living Theatre.

The least convincing of Cavani's features, *Cannibals*, trying to mediate between Creon's dystopia and the present quagmire of modern industrial Italy, only weakens the impact of both myth and reality.

In 1970 Cavani returned to television to make *L'Hospite* (*The Guest*) about the lives of the patients ('guests') in Pistoia mental hospital. The film is fictional, but makes use of documentary elements. It is primarily a celebration of the poetic aspects of madness, somewhat in the manner of Bolognini's *Down the Ancient Stairs* or Rossen's *Lilith*. The story concerns a writer who visits the hospital to do research and gets interested in a beautiful patient named Anna (Lucia Bose), who is recovering from a traumatic family death. This part of the film has something of the documentary inquiry about it, profiling patients and conveying information. In the second part, however, events become more personal. Anna is discharged and goes to stay with her brother. She is unhappy there and runs away to an empty villa which once belonged to an uncle. Here she withdraws and relives aspects of her childhood. Memory mingles with fantastic projections in the manner of classic love stories, Maeterlinck's *Pelleas and Mélisande* in particular (with musical accompaniment from Debussy's suite of the same title). Anna's dream-lover is a compound of her beloved dead cousin and a man she knows at the hospital, Luciano. The film ends with Anna's return to the hospital, and Luciano.

Tiso, in his monograph, describes *The Guest* (still unshown in the UK) as 'a work of fear, introspective, dark and cruel'. It was the showing of *The Guest* at the Venice Festival which provoked Bellochio, Fofi and others of the (then) Marxist film-making contingent to berate Cavani for her bourgeois position. It must have been all too apparent by then that the direction of their cinemas had diverged; Bellochio's seemed a cinema of despots, Cavani's of revelations.

Cavani's first feature to win her international status as a director, *The Night Porter*, came two years later. The film was produced by an American then living in Rome, Robert Gordon Edwards (who was also producer of

Beyond Evil four years later). *The Night Porter* returns Cavani to a period setting – Vienna in 1957. Vienna is a city which Cavani admires as much as her co-writer Moscati dislikes it. 'This is a city apparently untouched by conflicts,' he observed (*Letture*, April 1974). 'These conflicts are hidden under a coldness and obsession with conventions.' Cavani's fascination for Vienna might seem a little curious. It is known as the 'pensionopolis' of Europe, one in four of its population is over 60. The decrepitude of the city Cavani dramatizes via her principal location, an opulent hotel from the imperial era. Here the hero, Max, works as a night porter, providing curious services for the guests – boys for the old countess, spotlighting for Bert, who likes to dance semi-naked to *The Rite of Spring* in his bedroom, imagining no doubt that he is Nijinsky.

Max (Dirk Bogarde) is a former Nazi newsreel photographer now in hiding. His past is ambiguous. He certainly photographed atrocities in camps, as we see from the luridly coloured flashbacks, but did he participate? Max's past is thrown up at him with the arrival of Lucia Atherton, a beautiful waif whom Max was obsessed by when they met in a camp, where she was a prisoner and also Max's lover. Lucia is now very different, rich and married to an American conductor on a European tour.

The reunion is momentous. When Lucia's husband is ready to move on, Lucia invents a reason for staying, knowing she cannot live without Max.

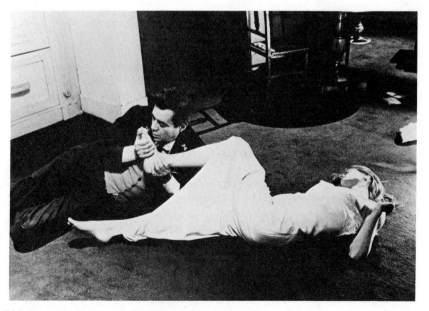

Dirk Bogarde and Charlotte Rampling in *The Night Porter*

Charlotte Rampling in *The Night Porter*

But Max's situation at the hotel is complicated, and Lucia, by staying, becomes involved in it. Max is one of several ex-Nazis now living in or around the hotel; their fanaticism, unlike his own, remains undiminished. Under the nominal leadership of Hans, now a psychiatrist, the group is involved in what amounts to a bizarre encounter situation, in which they enact a simulated war crimes trial, using a room at the hotel. The confession of their former crimes is not so much to cleanse their souls as to strengthen their group-solidarity. They are preparing, in fact, for the re-emergence of Nazism. The group is still ruthless; a stray witness from the past, as Hans suspects Lucia might be, must be rooted out and eliminated, lest he or she incriminate them.

Max is no less ruthless. He kills an ex-group member whom he knows once knew Lucia. His ruthlessness, however, is in the cause of his tragic attachment to Lucia. He gives up his job, hides Lucia in his flat, even chaining her to the bed so that she cannot be abducted. The group lays siege to the flat, blockading the groceries. Max and Lucia are starving. With their last ounce of energy, they smear dregs of jam on each other and make love one final time. They leave the flat and drive to a bridge over the Danube. It is dawn. They walk. Two shots ring out. They lie dead, like the Cannibals, their innocence returned to them.

The Night Porter earned for its director international recognition. She was helped by her cast, Dirk Bogarde and Charlotte Rampling, but she was also in command of an explosive storyline. The inspiration for the film came, Cavani has explained in detail in her introduction to the published screenplay (Einaudi 1974), from her research into camp survivors conducted during her period making documentaries for RAI in the early Sixties. In 1965 she had to make a television film called 'Women of the Resistance'. One woman told her a disconcerting story. After the war ended she returned every summer to Dachau, where she had been imprisoned for three years. Another survivor (from Auschwitz) was located living apart from her relatives in a shanty in the suburbs of Milan. She explained that she lived entirely alone because, 'You felt ashamed in front of others for having survived that hell, for being a living testimony to it.'

The film was also inspired by Nazi newsreel and propaganda material which Cavani had screened while preparing a documentary, *The Story of the Third Reich*, again for RAI, in 1962. 'The Germans loved to film,' she says. 'Hitler and Goering were great film buffs. My editor and I saw reel after reel of footage depicting the campaign on the Eastern Front ... the pictures of the twelfth century showing The Inferno were naïve in comparison. Evidently there'd been an increase in cruelty.'

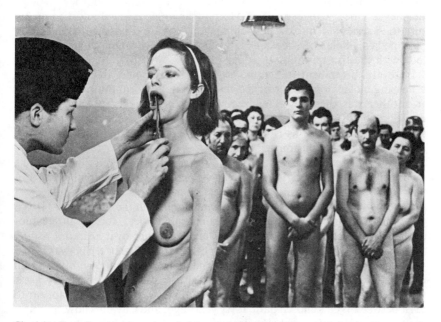

Charlotte Rampling seen here degraded but not yet abject in the concentration camp flashbacks of *The Night Porter*

67

Essentially *The Night Porter* reveals the penchant of Italian directors for stories which put a man and a woman alone in a room together in order to see what passions and lusts emerge. Cavani's film, at its heart, is *Last Tango* transposed in time and place, but sharing the same sexual delineation. Wertmuller's *Swept Away* and Ferreri's *Liza* also trap lovers together in remote places, and the latest in line is Montaldi's *Il Giocattolo* (*The Toy*), shown in Italy in 1979. In this, a husband and wife are besieged in their house following the husband's inadvertent involvement in a gang shooting. Montaldi's film ends with a less romantic vision of ruined love than Cavani's, however. The man tries to leave, his wife shoots him, without even removing the gun from the bedclothes.

Following *The Night Porter*, Cavani returned for a while to script-writing and television. Her 1974 tele-film, *Milarepa*, was made for RAI under the same production auspices made available to Fellini for *Clowns* and Bertolucci for *Spider's Stratagem*. The finance came from RAI but the film could be shown in the cinema simultaneously to transmission. *Milarepa* was based on the book about the Tibetan yogi, who was an 11th century poet, magus and hermit. As with *The Guest*, Cavani adds a fictional dimension. A young student today, Leo, and his professor, Bennett, are on the way to the airport to catch a plane to Tibet, where they are going to research the journeys of Milarepa. Their car crashes, not seriously, and Bennett's wife goes off to find help, leaving Bennett and Leo to begin their imaginary journey through mediaeval Tibet, with Bennett as Marpa (the yogic master) and Leo as his initiate. A series of consciousness-enhancing tests are conducted, as in Carlos Castaneda's similar mystical odysseys in Mexico with Don Juan. The film ends with the trio resuming their car journey.

In several prominent Italian *auteurs*, as already noted, there seems to be a 'missing movie', one that reached final screenplay form but for one reason or another was then never realized. In Cavani's career, too, the 'missing movie' is significant. In 1977 Einaudi published the screenplay she had written with Moscati and, as is clear from the lengthy preamble she supplied, researched with her customary diligence.

Lettere dall'Interno (*Letters from the Inside*), derived from Simone Weil's own writings and letters until her early death in 1943, presents us with a view of Weil that is faithful to the facts of her life as well as the shape of her thought, as was the case in *Galileo* and *Francis* in the mid-Sixties. It is a pity that Cavani never managed to realize this project, for, of all her biographical subjects, Simone Weil seems closest to her own attitudes to life. She was a French Jew, a young woman of great intelligence. She was also a political activist, working in factories and joining a combat unit in Spain during the period of the Civil War, though she did not display very great skill with firearms, as the script amusingly details. Certain parallels appear to exist between Weil's chosen path, forty years ago, and Cavani's now, attempting to realize her intractably intellectual productions in a film industry where very few women directors or writers exist.

The script begins by showing a conflict between Simone and the principal at the school where she teaches over her leftist political activities. He sacks her. We see Simone at home with her frightened middle-class mother, father and brother, who are uneasy about the future for European Jews, even in France, where they are living. We also see Simone with her fellow socialists, who are rapidly shown to be far more theatrical and who disapprove of her plans to work on a factory floor. She is separated from her socialist brothers by her religious conviction. It is this aspect of her personality which enables her to take a routine menial job, wash the dishes in a Spanish, bomb-wracked ditch, or sail back from sanctuary in America to join the French Resistance.

Rapid vignettes from work and leisure intersect. A worker, Ninette, advises Simone, 'If you want to get along, keep quiet.' At the opera house Simone perceives, 'To be a worker means not having the chance to appreciate beauty, therefore the condition of the worker is against nature.' This very direct, perhaps feminine perception, is exactly what set French intellectual leftists against her. To Vidal, a friend, the script shows, she sent her most passionately rebellious letters. She writes, for instance, 'I can feel the revolution in the tips of my fingers.' Wish-fulfilling? Perhaps. On the other hand, the political volatility of Europe in the Thirties could have produced a revolutionary situation.

Arrested by Fascists and interrogated, she avers, 'Fascism doesn't come from Mars. It's a disease inside Europe.' She escapes and hides briefly in a Benedictine Abbey, where she has curious conversations in Greek with the Abbot. On to New York, in 1942, feeling uneasy about leaving troubled Europe but torn between her sense of purpose and her family, now settled there well clear of trouble. She cannot stay. In London the Free French Resistance needs helpers. She takes lodgings in Bloomsbury and is put to work in London writing radio scripts for Radio France on the theme of deracination, a factor in her own life now which she feels acutely. Her observations are found by Cloison, the head of the department, to be 'sentimental and feminine'. The two of them clash frequently.

Ashford sanatorium. Flash forward: an old woman who works there, Carmen, recalls Simone's five months stay in 1943, before she died. 'She was wilful and obstinate like a child. She refused to eat, she wrote all the time to relatives in New York and made me post the letters from London.' Up until the end, in fact, Simone did not reveal to her family that she was dying.

In her last weeks Simone became aware of the decimation of the European Jewry and turned to old Sanskrit texts for consolation. 'These worlds are without joy because they are developed in darkness,' she read. At the moment of death she experienced a return of her earlier vision on the factory floor, in which her workmates downed tools and played Mozart together.

Carmen, whose presence concludes the film, has kept the last pages Simone wrote, and lines from these are incorporated in the final shots (of Carmen cleaning). Simone Weil's final self-assessment also seems to describe Cavani's own aspirations: 'I have sought with all my strength to free myself from the skin of the intellectual bourgeois.'

What they share essentially is a belief that fundamental social change demands first a change on 'the inside' – a different motivation for research, as Weil hinted in her essay on technology. The resonance between the thoughts of two European intellectual women with forty years standing between their lived experience is startling. *Letters from the Inside* is undoubtedly Cavani's most cogently developed screenplay and would, as a film, have compressed her quality of thought in one narrative in a way that has not so far happened in the realized movies. However, it should be noted that *Letters from the Inside*, too, has its moments of aberration. Simone visualizing the workers playing Mozart is a less plausible utopia than Fellini's mordant inversion, in *The Orchestra Rehearsal* – musicians disintegrating into loutishness. Cavani is also open to question when she insists on a parallel between Simone's radicalism in the Thirties and that of the mid-Sixties protest movement.

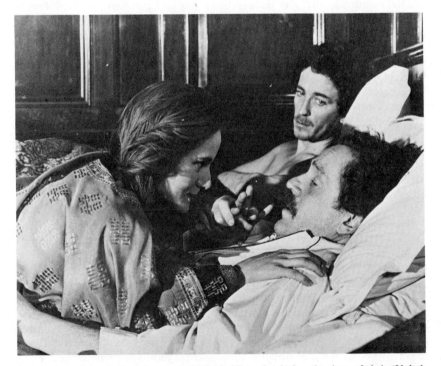

Lou Andreas Salomé, Paul Ree and Friedrich Nietzsche during the days of their 'Unholy Trinity' – *Beyond Evil*

Beyond Evil, of 1977, is divided into two distinct portions. The first deals with the coming together of Friedrich Nietzsche, Paul Ree and Lou Andreas Salomé in Rome in the early 1880s. They make the decision to form a *ménage à trois* which will be cerebral rather than physical. The effect of this *ménage* on their emotions and actions forms the final sequence of this section.

In the second part of the film we see the three members of the 'unholy trinity', as Nietzsche dubbed it, going their separate ways, though briefly Lou and Paul attempt to live together, moving from Leipzig to Berlin, while Nietzsche roams Europe, spending much of his time in Northern Italy until his mental collapse in 1889. The lives of the three characters are shown fragmenting, following the failure of the 'marriage'. Lou, half against her will, is impelled into a strange marriage with a young academic at Berlin university, Karl Andreas. Paul leaves the city and sets up a modest practice as a doctor in a remote town, where he treats the poor without much regard for making a living. Nietzsche becomes increasingly strange, drugging himself with laudanum and opium, submitting himself to intense hallucinations. The

... And Dominique Sanda, Robert Powell and Erland Josephson, who play the three disciples

71

film ends with the death of both men, and Lou's last-ditch attempts to come to personal terms with the permanent effects they have exerted on her life, both intellectually and emotionally.

The film, when shown in Paris in 1977, elicited some spirited objections from the critics. Jean Collet, for example (*J'Informe* 7th October 1977), accused Cavani of vulgarizing the material – 'She has preferred the sexshop to real history.' He recommended that she get some rapid psychoanalysis. The notion of Cavani's cinema as a 'sexshop' seems risible, given her always serious (often earnest) intentions. In London, in late 1979, when the film opened in a dubbed version under the shorter title, *Beyond Evil*, critical reaction was again generally hostile. 'Pretentious nonsense and a monumental mistake,' wrote David Castell (*Sunday Telegraph* 28th October 1979). 'Cavani's world is hollow, its characters unconvincing. Her Nietzsche is a ridiculous figure,' rasped Philip French (*Observer* 28th October 1979). Only Derek Malcolm managed to be a little more generous (*Guardian* 25th October). 'In this year of the bland masquerading as frisson, the film has genuinely eccentric guts.' Nobody mentioned Cavani's seriousness of purpose, first noted by Leo Pestelli in his introduction to the published screenplays *Francis* and *Galileo* (Gribaudi 1970) when he wrote that he admired her 'minimal effusiveness'. The accusation by French critics that she sought to make a meretricious movie, using Salomé/Nietzsche/Ree as mere erotic vehicles, evaporates in the face of Cavani's all too evident attachment to the implications of her study. When I talked to her in Rome in June 1979 she told me that she had read Nietzsche's work in its entirety during her research for the script (co-written with Italo Moscati and Franco Arcalli) and that the dialogue of the film was derived in the main from the correspondence which exists between the three protagonists.

All the same, Cavani's decision to throw some of the weight of the 'unholy marriage' back from the mind into the body was a critical decision. The scenario is acknowledged to be a 'free adaptation of real events' but all too often raises the vital question: at what point does freedom of artistic aim become distortion? From the documentary evidence there is no proof that Lou/Ree/Nietzsche ever conducted their 'marriage' on an erotic level. Reliable sources, such as Rudolph Binion, in his massively documented biography of Lou, *The Wayward Disciple* (Princeton University Press 1968), declares that Lou was a virgin not only at the time she married Andreas, but for a number of years after. In Cavani's film there is one scene in which Ree and Lou are shown making love, while Nietzsche expresses his uncertainty by first fondling them and then attacking Paul, at which point Lou evicts them both from the shared apartment with the curt dismissal, 'Intellectuals!' Elsewhere, during the pre-Leipzig period, Lou stays with Nietzsche at his mother's country home in Tautenberg, near the Black Forest, and here, too, Cavani implies strong erotic tension between them.

Accompanying this probable distortion of the sex life of the trio come other

Erland Josephson and Virna Lisi in *Beyond Good and Evil*

curious alterations in the dynamic of the 'marriage'. Nietzsche is presented as an amiable buffoon. Erland Josephson, in part, manages to catch some of the tenderness and secret chuckling of the man but fails (or is never given the opportunity) to provide any sense of his intellectual dignity, or grandiosity, depending upon one's disposition.

Cavani inflicts a similar degree of distortion on the Lou/Paul Ree relation-ship (although this, because less central, is perhaps more pardonable). Ree is played in the film by Robert Powell, with a sub-Shelleyan romanticism and anguish. Ree was in fact a short fat ugly man, very shy, rational to the point of desiccation, and strongly Jewish. It was Ree, incidentally, who minted the phrase 'beyond good and evil' in a short philosophical essay that was pub-lished at about the time he met Lou. Given Lou's celebrated attraction to older men of intellectual stature (towards the end of her life she became a close professional friend of Freud, and other eminent analysts within the so-called 'Vienna Circle'), Cavani's postulation of a romantic passion between Lou and Ree is implausible, though it is likely that Ree was infatuated with Lou, at least for a while.

Given this marked disorientation of dramatic emphasis – sex where there should have been none, and absence of any solid signs of intellectual development occurring in the three of them where there should have been much happening – the years 1881 to 1889 were by far Nietzsche's most brilliantly productive, while Lou also made a literary reputation during this phase – what is left to say about Cavani's 'adaptation'? She chooses to build her film around Lou Salomé, whom she presents to us as an intellectual bedevilled by two lechers. In Dominique Sanda's playing, Lou becomes imbued with a very French demureness, especially apparent when Nietzsche, drunk and drugged as usual, makes her do revolting things, such as making a modest micturation while he watches. Sanda has the natural composure to make even the act of urination look slightly *chic*; Lou Salomé was, however, a less well-mannered girl, by most accounts, who happened to come from St Petersburg not Paris. Cavani wants to show Lou as an example of a self-determining female. She wants to ignore Lou's all too obvious ruthlessness and, perhaps too, a certain intellectual opportunism. One of Lou's later lovers, the psychoanalyst Semek, described her dispassionately as 'the only truly evil woman I have encountered, though Lou's evil is of the kind which may bring good results'.

The ending itself contains a dizzy departure from the facts. Cavani shows Nietzsche insane, dressed in his white smock, now being tended at his family home by mother. This is shortly before he dies, in 1900. Lou, now a substantial literary figure, arrives in a coach with her current lover. She goes to the house and watches Nietzsche plunking the piano. In she goes, slipping past mother, engrossed in another room at *koffeeklatchen*. 'Our century has arrived,' she tells Nietzsche softly. His hands descend fervently on the piano. Mother rushes in. Lou is evicted for her Jewish taint. As she leaves she points out to mama that her son is playing 'Jew's music'. Off she rides in her coach with her handsome young lover.

Here one sees Cavani at her most characteristic. The scene is inventively imagined (Lou did in fact make an impromptu pilgrimage about this time in her life, but to Russia, to see Tolstoy, in company with Rilke) but pushes too hard for the momentous effect. It is also hard on Nietzsche to suggest that through his eleven years of mental attrition he had merely been in a kind of hiding, waiting to be regalvanized by the reappearance of Lou. There is some evidence developing from one Nietzsche scholar, Ronald Hayman, which suggests that Nietzsche's madness was in part feigned, more an abrogation of his famous 'will to power' than a state of utter confusion.

Is there any area of agreement in *Beyond Evil* between history and Cavani's presentation of events in and around the 'unholy trinity'? In the character of Elizabeth, Nietzsche's sister, some concurrence is reached, both in the casting and the scenic conception. Virna Lisi, in the role, looks both Teutonic and sullen. She lurks at Tautenberg while Lou and Nietzsche play cat and mouse with one another. On one occasion, waiting for them to return for lunch, she becomes so hysterically angry that she shreds her chicken with her fingers, to the maid's consternation. 'Anybody who could have seen us,' wrote Lou about her Tautenberg idyll with Nietzsche, 'would have thought that two devils were talking.' It is interesting that Lou's later perspective on those turbulent days is not so very different from Elizabeth's view of the relationship. Perhaps the much vaunted conflict between Lou and Elizabeth had its roots in a certain common fascination with the man's intelligence.

Cavani at present stands at a watershed in her career. She has made seven films, mainly for television, yet her critical and commercial status would seem to be waning. Two factors have mitigated against Cavani's kind of cinema in the past few years. The first is her asperity – 'My characters are lucid,' she told Tiso, 'and this makes them scandalous.' The second is her absence of belief in a Marxist political transformation. – 'It seems to me that only high finance benefits from a materialistic analysis of society,' she has stated. At the heart of Cavani's thinking, of course, there is a kind of conservatism. 'The modern world is an illusion,' she has commented. 'Over two thousand years the crisis points have presented identical characteristics.' This is not so different a perspective from Nietzsche's idea of Eternal Recurrence.

That Cavani has artistic faults we have noted already. Chief among these are her tendency to over-write, particularly in dialogue, and her visual reticence – she has none of that intuitive fluidity with the camera that one can

Dominique Sanda, Erland Josephson and Robert Powell in *Beyond Good and Evil*

point to in so many Italian film-makers. These faults, however, seem to me to be peripheral rather than central to her artistic identity.

Cavani herself has few illusions about her problems as an intellectual director who has to trade in the market-place. 'My audience thinks in a manner contrary to me', she has admitted. This may be a danger, for it implies a Simone Weil-like belief that society has no time for her kind of *virtuous* intelligence. Cavani's difficulties as a director seem finally more to do with her self-presentation than with the defects in her artistry. She needs to become more shrewd at judging how to entertain an audience.

In 1979 Cavani moved from cinema to opera production, directing a production of Berg's *Wozzeck* at the Maggio Musicale, in Florence. She showed her usual capacity for underlining the didactic elements to the drama. As Susan Gould, writing in *Opera* (November 1979) noted, 'Miss Cavani's intentions were to show the unfortunate figures oppressed and overwhelmed by society and the military world, and to do so through the inspiration of the music rather than the text, so freeing herself from too severe a tie to Berg's specific instructions'. The reviewer cites several illustrations of Cavani's cavalier treatment of the libretto, which she styles 'rather disturbing' but she concludes, 'I cannot deny the cinematic effectiveness of this Wozzeck'. It would seem, then, that the parallel with Visconti's artistic direction, already noted in the context of *The Night Porter*, is becoming more distinct in Cavani's career at this moment.

She remains a firm believer in the power of cinema. 'With the loss of independence of the press' she has said, 'with TV becoming increasingly the voice of the governing party, cinema has become the only space for an alternative. Cinema is hardly born.' One would like to share her optimism.

3. Form as Fatherscape: Mauro Bolognini and Bernardo Bertolucci

Curious intersections of theme and form occur in the cinemas of Mauro Bolognini and Bernardo Bertolucci. It is the purpose of this chapter to trace these through. On the surface there are marked divergences. Bolognini belongs to the generation of Fellini and Antonioni; he is known in Italy for his numerous movies made on unprestigious budgets for domestic market consumption (the tally today approaches some forty movies) but is less known outside Italy, although in France film critics are beginning to reappraise his achievement. A further apparent divergence is that of personal style: Bertolucci has always been a 'noisy' film-maker, from the late Sixties onwards, when he hectored European festival audiences about the revolutionary significance of *Partner*. Bolognini, on the other hand, works steadily and quietly. One reviewer aptly called him 'the cat that walks by himself' through the Italian film industry. Outside France and Italy he is usually regarded as somewhat lightweight. In *Film Dope* (No 4, March 1974) a reviewer noted his 'propensity for elaborate period reconstructions and his generally sure-footed way of crossing left-wing formulae with commercial ones', and added more callously, '... his films have always tended to be quickly forgotten, uncelebrated, gone to the graveyard every one.'

But Bolognini's cinema is better than such a crushing verdict allows. It is the aim of this chapter to show how two medium-weight *auteurs*, both of whose cinemas have been as afflicted with leftist ideology as they have been gifted with formal and expressive strengths, have undergone very different careers in the commercial arena, one suffering from the penalties of international inflation, the other the penalties of more localized deflation.

One area of affinity between Bolognini and Bertolucci lies in their common interest in Italian painting. This is ground they share with Pasolini, whose early films were composed in the manner of the Northern Italian renaissance masters, notably Mantegna. While Bolognini's cinematography and composition sometimes draws inspiration from the Italian impressionist painters, notably so in *Metello*, in which his shots of Florentine building sites seem to derive directly from Nomellini's Tuscan paintings, Bertolucci draws on other strands of twentieth-century European paintings beside Italian impressionism. Certain compositions in *The Spider's Stratagem*, for example, are contrived in the expressionist vein of Ligabu, whose paintings comprise the movie's credit sequence; similarly, in *Last Tango in Paris*, Bertolucci composes the fornications in the Plessy apartment with an eye on the wrecked figures of Francis Bacon's canvases (a major exhibition of Bacon's work was in fact showing while Bertolucci was shooting *Last Tango* in the city). Elsewhere in Bertolucci's cinema there are references to Italian impressionism, however. In *1900*, Pelizza's vast canvas *The Fourth Estate*, showing peasants on the march, makes up the film's opening shot, and the costumes of Pelizza's peasantry are closely observed by Bertolucci's costume designer, Gita Magrini.

The two *auteurs'* links with Pasolini extend beyond their congruous visual concerns, however. Three of Bolognini's early films stem from his period of collaboration with Pasolini. These are *La Notta Brava* (distributed in the UK under the title *Hot Nights*), *La Giornata Balorda* (*The Foolish Day*) and *Il Bell' Antonio* (*Beautiful Anthony*), made between 1959 and 1961 with scripts contributed by Pasolini. Bertolucci's debut feature film, *La Commare Secca* (*The Grim Reaper*), made in 1962 when Bertolucci was only 21, also derived from a screen original by Pasolini, who was a friend of Bertolucci's father, Attilio, a film critic and poet. Pasolini had already encouraged the young Bernardo's efforts at poetry, recommending a publisher for his first volume, *In Search of Mystery*, and using him as dialogue director on his own debut feature, *Accattone*, in 1961.

In fact, when Pasolini shot *Accattone*, he covered similar themes to those he had developed in the scripts for Bolognini – the same stress on lower depths life among pimps, whores and hustlers of the outer city *barracci* – while imposing on his filmic terrain a harsher, dustier ethos.

The bravura of these early Bolognini films is epitomized by the opening shot of *La Giornata Balorda* – a slow tracking shot through a teeming tenement courtyard which is centred on a massive line of washing. As the camera approaches the washing random voices are heard calling from balconies, mingled with natural sound effects. Directly below the washing-line the camera pauses. Voices now separate into distinct lines of dialogue. Women are asking about a man who is missing. As the camera emerges into the light beyond the washing-line a reversed perspective takes place. We see a dustman traipsing the balconies collecting the rubbish and talking to the women. The grievances about the missing Nino continue. His young wife is pregnant; Nino is jobless but, instead of looking for work, he sleeps all day, or simply disappears.

What Bertolucci and Bolognini share, formally speaking, is a concern for an expressive *mise en scène* which mirrors the emotion of characters. Not only the depth of field and the quality of light counts (though in these respects Bolognini, working most notably with Nanuzzi, and Bertolucci, working with Storaro, have achieved mellow painterly qualities) but also the dynamics of

La Notte Brava

cinema composition. The track, the tilt and the crane are vital elements.

Style apart, what links Bolognini's *La Notte Brava* and *La Giornata Balorda* with Pasolini's *Accattone* and Bertolucci's *La Commare Secca* is the common theme of life lived at the 'dead level'. Among the street apaches and prostitutes of these films a sense of honour is everything, for life is lived without expectation. Bolognini told Jean Gili that what he and Pasolini shared above all was a sense of despair at the conditions in which these unemployable dwellers in the dusty, hovel-strewn Roman *periferia* are forced to live. This sense is distinctly revealed in the character of Accattone himself, forced to steal his own infant son's necklace, muttering 'may God forgive me' as he does so, and also in Ruggero, one of three apaches in *La Notte Brava*, who lives dangerously, doing a bit of pimping, stealing and gun-running.

There is one inspired scene in *La Notte Brava* showing how the three hoodlums react to the three cruising whores they pick up on the road between Rome and Fiumecino, where they have been trading some stolen weapons. The three couples wander into open country with spinneys and a dyke on which are sprinkler hoses. Separately, Bolognini observes the behaviour of each 'couple'. Bella-Bella, the slowest-witted of the trio, takes his girl Nicoletta down to a swampy dell and gives her a line of brag which is clearly extraneous to his immediate purposes. Scintillone runs down a bank with his girl, Anna, without ceremony, and grapples with her.

Ruggero, however, rises to other possibilities contained in the moment. They must make it lyrical if they can, he senses. He and his girl, Supplicia, walk along the dyke under the sprinkler hoses. They stand, scarcely touching, wet, in moonlight. They have seized and particularized their moment, and achieved a certain minimal affection.

The differentiation between Ruggero and the other two gangsters is developed in a sequence occurring slightly later, which shows them returning to a mansion owned by several rich homosexuals who have accosted them. Bella-Bella and Scintillone, installed on settees with large drinks, resort to pandering. The Marxist connection is here explicit: rich boys and poor boys juxtaposed in the only common activity society makes available to them – debauchery. But Ruggero's role again transcends this erotic opportunism. He wanders off and meets a girl, Laura, sister to one of the homosexual youths. They make love. As Ruggero leaves with his cronies he finds that they have stolen money. To Ruggero, this seems dishonourable, even with his none too fastidious standards. He and Bella-Bella fight, but meanwhile Scintillone makes off with the money, meets an old flame named Rosanna and takes her out on the town. They are evicted from a club for drunkenness, just as Ruggero arrives. Scintillone is hauled off by the police and Ruggero steps into his role with Rosanna. They dine and dance in a restaurant. On the way home Ruggero stops the cab and picks dog-roses from the hedge for Rosanna. He drops her home, pays off the cab, and finds one last banknote in his pocket. Leaning, replete, over a bridge, he tosses it down. The camera closes in on the crumpled banknote lying in the garbage. This shot was also, we now realize, the opening shot. The cyclical device is a favourite item in Bolognini's visual repertoire.

Ruggero, like Nino in *La Giornata Balorda*, lives without scruples, using his wits, yet is oddly undepraved. Nino's search for a job reveals the extent of Roman nepotism and corruption. Inadvertently involved in a *Cosa Nostra* racket, he tells the mobster boss's moll who seduces him, 'I want to live with you forever.' At the end of the day, the still jobless hero returns to his girl and young son in the crowded tenement. He gives a stolen ring to his girl and holds his sleeping child close. 'Grow up soon so that·you can go dancing with your father,' he mutters. His girl comes up behind him. She understands how depressed and helpless he feels. Bolognini ends where he started, following this haunting moment of family intimacy with a lugubrious track down the tenement yard, which now seems like the hall of Hades. Can the youth's vitality hold? Will he still want to dance when his son is twenty?

Bertolucci's debut feature, *La Commare Secca* (literal meaning 'the dry housewife' – colloquially *The Grim Reaper*), covers similar ground to Bolognini's *La Notte Brava* and Pasolini's *Accattone*, showing the milieu of Roman back-streets, the banks of the Tiber and the *autostrade*, the shabby tenement courtyards and apartments, and the pimps, whores, street hustlers and manual workers who populate this world. Yet there are also differences of emphasis. For Bolognini and Pasolini the dramatic questions are social: will Accattone survive as a street apache? Will Ruggero (*La Notte Brava*) get caught by the law? Will Nino (*La Giornata Balorda*) find a job?

Bertolucci, however, chooses to dramatize his back-street milieu more starkly, in the form of a police investigation following the death of a prostitute down by the Tiber. Four suspects are questioned about their whereabouts on the night of her death. Using a *Rashomon*-like device – a grid of subjective flashbacks depicting each man's activities during a thunderstorm which occurred just before the murder – Bertolucci is able to present a sharply polarized view of innocence and depravity, while including brief shots of the victim herself alone in her room during the storm to serve as a refrain or bridge passage.

Two of the suspects are nefarious – Califfo the pimp, and Canticchia, the sneak-thief. Both exploit human need for sex. Califfo, faced with a whore who cannot render his quota, takes her dog. But his partner, a tough girl called Esperia, is more than a match for him. When he tries to leave her, she lies in wait for him in the Paolino Park (where all these flashbacks take place) with a knife. Canticchia is equally cowardly. He is shown with a gang trying to steal a radio from a pair of lovers on the edge of the park.

The Killer apprehended in *The Grim Reaper*

He is caught. 'Feeding off people in love!' the outraged girl shrieks at him. Canticchia tries the familiar bluster – 'I'm a poor orphan,' he pleads, making a run for it.

Balanced against these 'songs of experience' are the innocent interludes concerning the soldier, Teodoro, on leave in the strange city and looking for a woman, and the exploits of two youths named Franco and Pipito, seen with their giggling girlfriends first at the zoo, and then at the apartment of an older sister of one of the girls, where they dance to a record player and plan a home-cooked lunch together, which will be dependent upon the boys raising some funds to pay for the menu. These elements have a freshness and exuberance which is more reminiscent of Olmi's *Il Posto* than Pasolini's already soured vision of sexual exploitation. Two scenes in particular point to the young Bertolucci's virtuosity: one shows Teodoro following girls in the street, covered by a nervily panning camera, a scene which is beautifully composed (and vividly edited by Nina Baragli) into a climax of feminine flurrying.

Equally effective (though for the tenderness of observation rather than the wit) are the two vignettes showing Teodoro alone in Rome. First, as he wanders among American tourists at the Coliseum, a mere sightseer himself; and second, when he shelters during the storm, and invites a bunch of bedraggled whores to share his refuge. The gleam of satisfaction on his face as they squelch into the tunnel, and the camera discreetly tracks away into daylight, is as delicately judged as anything in Bertolucci's better known second feature, *Before the Revolution*.

Considering the film in the *Guardian* (9th November 1962), Richard de la Mare and Maurice Hatton observed, 'The flashbacks are about to become monotonous when Bertolucci reverses the whole process – we see the result of a cross-questioning. We see the murder, we know who the murderer is, but apparently he has deceived the police and got away. Then ... we see the arrest ... the carefully constructed cardhouse of question and flashback is flattened by a straightforward arrest.' It should merely be added that the presentation of this arrest is realized with a bravura that is later to become Bertolucci's hallmark. First, a general shot of the dancehall, then the face of the homosexual 'nark', with the sound of Natalino's wooden clogs impinging before the camera isolates him dancing. Here, as elsewhere in this intricately organized film, sound and picture work as distinct elements, serving to show the early technical confidence of Bertolucci.

In 1960, Bolognini shifted his attention from the Roman backstreets to the middle-class boulevards of Palermo in Sicily. The young hero of *Il Bell' Antonio*, Antonio (played by Marcello Mastroianni), returns home to his parents' apartment after an unsuccessful attempt to make good in Rome, where he has failed (so the opening scene shows) to cope with either work or women. His father is a bombastic businessman whose favourite boast concerns the night he had nine women. He is anxious to know whether the local rumour – that Antonio has cuckolded a powerful Roman politician – is true.

Antonio's crisis of identity manifests as impotence but the root of his problem is a crippling narcissism. The girl he marries he dotes upon, calling her his 'angel'. But the relation remains unconsummated. 'Sometimes a man can love too much,' remarks his doctor, mournfully. Antonio and his wife, freed from economic exigency by the income settled upon them by both conniving fathers, take up an existence in a villa in the country which amounts to a form of banishment. Faced with extracting pictorial equivalents for this parlous situation as presented in Brancati's novel, Pasolini and Bolognini choose to depict the young couple in a series of scenes suggesting emotional arrest. We see Barbara on a swing, while Antonio gazes adoringly at her. Already she is irritated with him.

Coupled with this unease is her sense that all is not right with the physical aspects of her marriage. Another tableau shows Barbara by a tree with her maid, asking how a child is born. The maid explains. Barbara runs off in horror. She is still intact. Drama descends on the household when she informs her parents. Puglisi, Barbara's father, insists that the marriage is nullified by the fact of non-consummation. Antonio's father, mortified at this blow to the family honour, takes a whore to bed to prove that he, at least, is a man, and suffers a fatal heart attack.

Linked to the inter-familial consternation is the theme of male friendship, as particularized in Antonio's relations with his cousin, Eduardo, who has also returned from Rome sadly disappointed. 'I've changed. I feel disgust,' he tells Antonio, visiting him at his sickbed. 'I got to know Italians. It makes me afraid, now I understand why you left Rome.' Eduardo also shares his cousin's disturbed attitudes towards women. Antonio's reputation as a man is vindicated, however, when a servant girl discovers she is pregnant by him. In the final scene Antonio receives a congratulatory phone call from Eduardo. 'Why should I be happy?' replies Antonio. 'I'm crying for Barbara.' In the final shot, as the camera creeps in, he weeps and murmurs, 'I want to be like everyone else.' The image freezes on a superimposition of the wall behind him – a characteristic softening of focus which Nanuzzi seems to favour when shooting for Bolognini.

Il Bell' Antonio is a film of great visual assurance, Bolognini using the beauty of the young Mastroianni's face as a pure composition. Male narcissim, female pragmatism, these are the motifs which constitute the emotional tension, with a touch of malice, perhaps, in the reiterated image and dialogue concerning Beautiful Anthony, the love object. 'He has eyelashes like fans,' shrieks one coquettish neighbour. 'Kiss him for me.' The predatory sexuality of Southern Italian women is an undercurrent to which Bolognini responds with relish.

In 1961 Bolognini adapted for the screen a Mario Pratesi novel, *L'Eredita*, with Vasco Pratolini, Pasquale Campanile and Massimo Franciosa as the scriptwriters. The film, retitled *La Viaccia*, was set in Florence of the 1880s and, like Bolognini's later film, *The Ferramonti Legacy* of 1976, takes as its theme a family feud deriving from a disputed legacy. A consistent motif in Bolognini's cinema has been his dislike of the rapacious owner-classes.

In this, of course, he is one of a number of Italian *cinéastes*, but his strength in this context derives from his ability to penetrate the texture of bourgeois life – the intrigue, avarice and back-biting – rather than assuming a lofty position, dependent on dogmatic ideology or outrage.

Bolognini is a Florentine and although he has shown considerable versatility of approach in handling other Italian cities as his background, not only the Rome of the early films but also Trieste, where *Agostino* is set, and Sicily, where *Il Bell' Antonio* and *A Beautiful November* are located, his films set in Florence seem to have a particular visual care. *La Viaccia* concerns the defection from the family farm of a renegade son, Amerigo, played by Jean-Paul Belmondo, who takes a job in his uncle's pharmacy in Florence, only to fall in love with a whore, Bianca, who works in the local bordello. He steals money from the shop to pay for his nights of pleasure, is detected as a thief, thrown out, and publicly beaten by his father. After a brief flirtation with a shadowy anarchist, who wants him to join the cause and leave the city, he returns to the brothel to take up duties as a bouncer, in order to be close to Bianca. Wounded in a brawl with another patron he limps home at night to La Viaccia, to arrive in the middle of yet another family quarrel concerning the disposition of chattels. He collapses outside the gate. 'Remember, brains and bludgeons,' the dying patriarch of La Viaccia had told him. Amerigo's tragedy is that he has neither the wit nor the aggression to hold onto life. He is a doomed creature in a milieu given over to dog eat dog aggrandizement for which he has no appetite himself, yet no alternative to head for.

Bolognini's theme – the difficulty experienced by bourgeois youths who feel no affinity of aim with the greed of their parents – is taken up by Bertolucci in his second feature film (also developed without the benefit of a Pasolini script), *Prima La Revoluzione* (*Before the Revolution*), made in 1964. Fabrizio, the hero, is a young man still living in the family home in Parma. His main diversions are theoretical politics – animated arguments with his peers as they walk the banks of the River Po – and watching old movies in the local cinema. At the beginning of the film, two events occur which interrupt his routine life. A friend is found mysteriously drowned in the Po; and Fabrizio's beautiful aunt, Gina, comes to stay. Her exotic presence, in the days of grief, leads to an affair. Finally, Gina leaves him, and Fabrizio contemplates marriage to a childhood sweetheart.

Fabrizio has one significant relationship, with an older man, Cesare, a quietly committed Marxist schoolteacher. This relationship, though peripheral to the narrative, is essential to the film's meaning, as are equivalent relationships in Bertolucci's later films – Jacob and Petrushka in *Partner*, Athos Mangiani and his dead father in *The Spider's Stratagem*, and Marcello Clerici and Professor Quadri in *The Conformist*. Cesare serves as a political conscience for Fabrizio; and, after he is plunged into the turmoil of an affair with Gina, as a calming presence. Gina, on the other hand, suggests the romantic inspiration to Fabrizio. This axis formed by the friend and the lover

Francesco Marilli in *Before the Revolution*

is echoed by the geographical axis of city and province. Cesare belongs to the quiet land around Parma; Gina comes from Milan. 'What do you do all day in Milan?' Fabrizio asks her during the funeral of his friend. 'I bathe three times a day. I laugh. I cry,' says Gina.

Cesare is Fabrizio's 'Garibaldi', so Cesare tells Gina when Fabrizio introduces them to one another. Gina feels jealous of this special relationship between the two men. The long scene in which Gina and Cesare tussle for emotional possession of the youth is one of the most skilfully rendered in the movie, and suggests a flash of maturity in Bertolucci's perceptions – he was only twenty-three when he made the film. When Cesare talks of the impossibility of political change and the myopia of the Italian people, Gina mocks him with a quotation from Oscar Wilde – 'the man of action is the greatest dreamer.' Cesare retaliates by suggesting that Gina's current hold over Fabrizio is manipulative, and therefore potentially destructive – 'You can receive everything and give nothing,' he tells her. As Gina leaves with Fabrizio, Cesare says quietly, 'The cruellest thing is to disbelieve someone's sorrow.' This is a reference to the effect the death of Fabrizio's friend has had on him.

(L to R) Francesco Marilli and Morando Morandini in *Before the Revolution*

This scene marks the slack water in the affair. After, the tide has turned. Gina understands that she must leave Fabrizio. She picks up a man in town, and makes sure that Fabrizio sees her do it. Fabrizio accosts them, then turns away, hurt and rejected. But so, too, does the prospective pick-up, sensing perhaps that he has served some ulterior purpose on Gina's part. Gina is left alone, sharply focused as the two men blur into the background. The visual comment is a classic Bertoluccian device; he makes similar use of deep field and switch of focus in *The Spider's Stratagem*, notably in the scenes between Athos and Draifa, the mysterious widow.

In the final scene between Gina and Fabrizio before she leaves, Gina says, 'I hate men. With their women, their families. I like you because you're not a man yet.' But Fabrizio has changed. This is made clear in the scene when he stands with a landowner on the banks of the Po, listening to the squire lamenting the closure of his estates – 'They will come with dredgers. Who will protect the poplars from the frost? No more pike, ducks, geese. Here life ends and survival begins.' Over this speech the camera rises lyrically above the river and poplar groves. Fabrizio, however, reacts sharply to Puck, the land-owner. 'It's easy to examine your conscience when you're broke,' he declares, acutely. 'There is no escape for the children of the bourgeoisie.' This remark seems central to Bertolucci's subsequent development. It synthesizes the difficulty: how to choose between political conscience – for the majority life has *always* been no more than a matter of survival – and the weight of one's own cultural upbringing, out of which one mourns with Puck the passing of a state of natural balance.

On this elegiac note Fabrizio's summer ends, and Gina goes back to the city. Fabrizio talks to Cesare in the town square – 'I'm like a pigeon in the square,' he comments bitterly. 'In twenty years the people haven't learned a spark of conscience.' Cesare upbraids him for his loss of idealism. 'What's wrong with workers wanting to better their conditions?' he asks, as a flag parade begins. 'I want a new sort of man,' says Fabrizio, 'one wise enough to educate his parents.' For Fabrizio, as for subsequent Bertolucci heroes, radical politics becomes a form of parent-bashing.

That winter, though, Fabrizio runs into Gina by chance in the opera house. He is there with his fiancée, a local girl. He slips out to talk to Gina, to a background of Verdi. Gina says she likes to attend opera in Parma because 'here the people believe in it'. Fabrizio tells her that his fiancée is 'not much but she's what I want at the moment'. Their mutual cynicism is the bond which now unites them. Fabrizio will soon get married. Another rebel bites the dust, the choirboys' grinning faces at the ceremony signal.

Many of the emotional tensions as well as the intellectual doubts of *Before the Revolution* are reiterated in Bolognini's *A Beautiful November*, made four years later in Sicily, from a novel by Ercole Patti. Both films study the incestuous relationship between aunt and nephew and end with the youth marrying on the rebound. The difference in region only serves to underline the essential similarity for both deploy scenes of erotic passion with bravura *mise en scène* in the landscape: Fabrizio's interlude with Puck on the Po is echoed in *A Beautiful November* by a shooting scene in the woods, in which Cettina, the youth Nino's aunt, demonstrates her marksmanship among the shotgun-toting menfolk. Cettina, like Gina, serves as a vital catalyst in Nino's adolescence. She, too, is a displaced person, with no clear role to play in a family whose structure is still patriarchal. She is neither child-bearer nor careerist, for her husband is rich and tyrannical. Like Gina, she feels able to define herself only via sexual adventures.

Although Bolognini has said that he feels less than satisfied with *A Beautiful November* – he failed adequately to define the character of Cettina, in his view – his handling of character in this film is distinguished by moments of subtle observation. To show the degree to which Cettina is subjugated within a milieu of male values, Bolognini and his four scriptwriters evolve a secondary sexual liaison, which takes place between Cettina and a business associate of her husband's, a bachelor named Sasa. In the first scene in which Cettina meets Sasa with her husband Biago, Sasa makes his admiration obvious and Biagio comments with satisfaction, 'Beautiful possessions are to be admired by others.' Later, in scenes between the three of them, it is evident that Biagio is using his wife as a means of involving Sasa in a business alliance. He encourages them in their mutual attraction, thereby accelerating the affair, which begins in the woods, in an old hunting lodge. For Nino, however, now infatuated with the aunt who has become his lover, her new liaison proves traumatic.

A further parallel with *Before the Revolution* can be discerned in Nino's friendship with a slightly older cousin of his, Umberto, who is shown to be more in touch with himself and shrewder about the measure of debilitation all around them. When Umberto notices Nino's confusion over Cettina he advises him to leave rapidly for London, and a new life. Nino, however, is confident that the corruption of the clan will never affect him. 'We'll never be like that,' Nino boasts to Umberto as they stand apart from the family's religious rituals at the rear of the family chapel. In the end, of course, it is Nino who turns out to be just 'like that' while Umberto, who has learnt the necessity of maintaining his distance from tribal corruption, seems set to make the vital break. The final scene shows Nino getting married to a childhood sweetheart while the tribe coos *Bravo* in predatory close shots. The circle has closed around Nino, as it did around Fabrizio. As in the Bertolucci movie, there is a coda, depicting Nino slipping aside from his bride to whisper conspiratorially with Cettina. The sexual complicity between them, far from ending with Nino's marriage, is just beginning. Nino is now prepared to use his marriage as a screen to provide him with a minimal equilibrium from which to launch his depredations: he is revealed as no different after all.

Formally, *A Beautiful November* shares with *Before the Revolution* great visual lyricism. The opening sequence, showing Nino and his small brother dodging in and out of a Church procession through the streets of Palermo, is a brilliant evocation of the power of the Church, its baroque facades, the heaving bearers below the vast painted ikons, the fireworks. The two bright-eyed nimble boys skipping among these weighty props of Sicily's first super-structure are sad to see; soon their vitality and innocence will be vanquished. Another masterly moment of *mise en scène* takes place in the opera house, within a box where Cettina and her sister sit, with Biagio between them. The sisters vie for control of the boy they both love, one as a mother, the other as a lover, and Bolognini handles this essentially feminine form of competition with intricate control of angle, movement, gesture and editing rhythm. This capacity to depict an entirely female dynamism is a hallmark of his skill and sensitivity.

1964–8 were fallow years for Bertolucci, who marked time writing a western for Sergio Leone and making an oil documentary for RAI television. In 1968 came the film he now repudiates, *Partner* (still not shown in the USA). *Partner* was loosely based on Dostoievsky's story *The Double* but shows strong traces of two contemporary influences on Bertolucci. First, he collaborated on the script with Gianni Amico, whose own film *Tropics* was also presented to European festival audiences in 1969. It proved no less didactic in purpose than *Partner*, being a semi-documentary about the plight of destitute Brazilian peasants on a trek to Sao Paulo. *Partner* prompted James Price, reviewing the film in *London Magazine* (February 1969), to tender the pertinent thought, 'Some of us have read less than others, the question

Pierre Clementi in *Partner*

is, does it have an identity of its own?'

The answer is, not one identity but many, not least the identity of Amico, with whom Bertolucci later admitted (Jean Gili interview 1977) to sharing 'a cinephilic delirium. We had the impression of risk but at the same time we sheltered behind pretexts of rigour and experimentation to justify our refusal to face up to communication.' This is certainly apposite.

But *Partner* was afflicted with another more immediately recognizable 'delirium' – that of Julian Beck's anti-aesthetic of Living Theatre, whose representative (yet again) is Pierre Clementi. Clementi, as Jacob, struts, rants and wheedles his way through almost every scene in the movie, playing two roles, Jacob I the art teacher, and Jacob II his messianic alter ego. Jacob II preaches guerrilla tactics on the streets of Rome, teaches his students how to make a Molotov cocktail, and generally commits, in badly differentiated fantasy vignettes, those violent acts which Jacob I dare not commit – getting arrested and attempting suicide in a urinal.

Of course, it's all a joke, in the Antonin Artaud manner. Mussolini rides by on a horse, Jacob dances in the street with his enormous shadow and pours his tea from a cow-shaped milkjug, but the humour is never very funny.

Jacob wants people to 'burst with anger' (a more vehement version of Fabrizio's remark about wise children having to educate their parents). But when Jacob I comes in from his girlfriend's, he still sees himself as Ulysses returning from Ithaca. The rebel remains a literature graduate, like his author.

'I was sick from the neurosis of style when I made *Partner*,' Bertolucci told me in 1978. Two years earlier, he told Jean Gili that the production was damaged by the social turbulence of the period, pointing out that in 1968 Rome University was occupied by leftist students for several months. 'In 1968 people were saying marvellous idiotic things,' he explained, 'like "the camera is a machine-gun". Me, too, a bit, I was caught up in the idea. In fact, it's a romantic idea, infantile and blind.'

Is *Partner*, then, also idiotic and blind? Perhaps. What matters, though, is not the terms of a denunciation but a consideration of how *Partner* reveals Bertolucci's limitations. There is in the film, for instance, a curious anti-sexuality. Love is polluted with slogans. 'Bed is the most dangerous place,' says Jacob as he frolics unplayfully with a detergent salesgirl (Tina Aumont, looking bewildered). Jacob's treatment of his regular girl, a member of the owner class and therefore an 'enemy', is no less uncouth. In one scene he rolls her down the hill in a car without an engine and proceeds to rape her, watched by his retainer, Petrushka, whose function is to out-vie Jacob's anti-aesthetic. 'Cinema cannot compete with spectacle,' shouts Jacob. 'In a world where everyone is an actor, the theatre is redundant,' ripostes Petrushka. With hindsight, we are able to understand that what Bertolucci in 1968 appeared to be celebrating was in fact the very process of social deterioration which made him desperate. What *Partner* suggests to us now is that Bertolucci is a spiritual conservative, better suited to mourning passing orders and receding forms of social stability than he is at announcing the dawn of a brave new world. In the end, the *putsch* in Rome which Jacob II attempts to provoke, via his dissident students, fails to happen. A power station is stormed, but not commandeered. Jacob climbs onto a high window-ledge, along with his perfidious partner. 'They want you to jump because they hate you,' Jacob II whispers.

The Fascist era in Italy was a time of cultural, political and emotional suspension. It was marked by a great civic and cultural hypermania in Rome itself and equally great isolation from outside influences in Europe which might have allayed the Mussolinian programme. In the film industry, as Francesco Savio's monumental study of Fascist movies, *Cinecitta anni trenta* (Bulzoni 1979), makes clear, 722 feature films alone were made between 1930 and 1943, yet few of them were important and importation of foreign films was forbidden. As Masolino D'Amico has pointed out (*Times Literary Supplement*, 15th February 1980), 'Censorship tended to be subterranean ... much of what was forbidden was later to become distinctive of neo-realism.'

Bertolucci first engaged the Fascist theme in 1969/70, in his film for RAI television *La Stratagemma dell Ragno* (*The Spider's Stratagem*), and followed it shortly afterwards with *Il Conformista* (*The Conformist*) in 1970. After the

Pierre Clementi and Tina Aumont in *Partner*

sub-Godardian excesses of *Partner* these two films mark his attempt to go 'straight' (and also perhaps to get himself 'straight', for he began an extended period of Freudian psychoanalysis at this time). Yet both films are stylistically distinctive; they stray far from the realistic depiction of fascism as portrayed in, say, Leto's *Black Holiday* or Scola's *A Special Day*.

The Spider's Stratagem, based on a short story by Borges, *Theme of the Traitor and Hero*, marks the moment when Bertolucci declared a moratorium upon his current political preoccupations and turned inward. The film marks the onset of a period of imaginative recall. In the film, a young man named Athos Magnani returns to his home-town – called Tara in the film but actually Sabbioneta, near Mantua – to try and find out what happened to his father, also named Athos (the same actor, Giuglio Brogli, plays both roles). Athos Senior has a reputation in town for being an anti-fascist hero and martyr. His memorial plaque stands in the town square with the inscription, 'cruelly murdered by Fascists'.

In Tara, Athos becomes involved with various citizens who once knew his father, three of them now elderly men and one an enigmatic woman, Draifa, who was his father's mistress and still lives in a crepuscular villa on the edge of town, looking no older than she must have done in the mid-Thirties.

Giulio Brogli (R) and Pippo Campanini in *The Spider's Stratagem*

It is from Draifa that Athos learns of his father's assassination by gunshot in the local opera house during a performance of Verdi's *Rigoletto*. From Draifa, Athos picks up random details about his father's character, not least his famous sense of humour. In flashback we see Athos Senior with his cronies on a country road at dawn, crowing like a cock to rouse the neighbouring roosters. 'He liked to joke,' says Draifa. But she also admits that she has not slept since he died. This remark, in the light of subsequent events, becomes significant. Nobody sleeps in Tara, and nobody grows older. Even stranger, there are no children, except for one odd, androgynous child who is apparently living at Draifa's villa.

It becomes plain, then, that Bertolucci's intentions are metaphysical not social. He examines the impact of fascism from a recondite perspective, suggesting that nothing has died since 1936, when Athos Senior was murdered. All those events live on. This bias towards an idea of Relativity is derived partly from Borges himself, whose recurrent theme is the circularity of time, and partly from Bertolucci's collaboration on the script with Eduarde de Gregorio, whose own subsequent films as a director have pursued the same theme of Relativity – *Sérail* for example.

Much has been made by critics of the father/son motif which manifests in *The Spider's Stratagem* and *The Conformist*. Richard Roud, for instance, examined this theme in an article in *Sight and Sound* (Spring 1971), and cites Bertolucci's own remarks on this theme, to wit – 'My father was a poet and I wanted to compete with him ... I used to write poetry myself, but I realized I would lose that battle, so I had to find a different terrain on which to compete.' His alternative terrain, the cinema, was of course a common ground between Bernardo and Attilio Bertolucci, for his father was a film critic on the Parma Gazette and had helped inspire Bernardo's early passion for the cinema, notably certain Hollywood directors like Stroheim and Ophuls. Bertolucci's attitude towards his highly cultivated family background was for a while ambivalent. Once he described it as one of 'suffocating culture'. Very recently, however, in a post-*Luna* interview (*Film Quarterly*, February 1980), he has commented more charitably on the family, describing his father as 'a lovely man' and hinting, for the first time, at a childhood that was equable.

It seems probable that Bertolucci's remarkable vitality as film-maker can be traced back to a happy childhood. This vitality distinguishes his cinema. Intellectually speaking, his cinema seems less coherent than, say, Liliana Cavani's, and less politically integrated than, say, Marco Bellochio's. But in the sheer verve of his *mise en scène*, he remains a master.

It was in 1970, with his adaptation of Moravia's major novel *The Conformist*, that Bertolucci's stunning technical vivacity became clearly apparent. American finance, from Paramount, must have contributed to the technical resources of the production, but it should also be noted that Bertolucci worked with Franco Arcalli for the first time. He claimed, when I talked to him in 1978, that from Arcalli he developed a new respect for the importance of editing. 'Before, I saw editing as an imperialistic process, a manipulation.' The editing is highly 'imperialistic' in *The Conformist*, tight and nervy, and matched by the compositional versatility which, in its use of diagonal deep field and wide angle shots, makes Il Duce's Rome resemble an imaginary city far more intimidating than Godard's Alphaville. The collaboration with Arcalli extended beyond *The Conformist*, however. *Last Tango in Paris* was co-written with him, and so, too, was *1900* (with further help from Bertolucci's younger brother Giuseppe). In 1978, during the scripting treatment of *La Luna*, Arcalli died suddenly.

The period of *The Conformist* is 1938, the setting is Rome and Paris. The drama concerns the assassination of an ex-patriate academic, Professor Quadri, who has defected to France to escape the restrictions imposed on Italian intellectuals by Mussolini. In Paris he is contacted by a former student of his from Rome University. Marcello Clerici (Jean-Louis Trintigant), who is ostensibly in Paris on his honeymoon but is in fact a Fascist agent here to set up the assassination of his former mentor. The killing does eventually occur –

with an operatic ferocity reminiscent of Julius Caesar's murder in the Senate – Quadri and his beautiful wife Anna (Dominique Sanda) are ambushed in snowy woods as they drive to Savoy, Quadri is stabbed to death by Fascist thugs and Anna is gunned down in the forest while Clerici watches. The death of Anna had not been premeditated, however. Clerici has become infatuated with her. Anna's last-minute decision to accompany her husband out of Paris has been her downfall. This is the gist of the drama itself, yet both novel and film make use of a wider range of material than is strictly necessary for the purposes of a political thriller.

Clerici's character is central to the meaning of the book and movie. He is a man obsessed with the need to pass as normal. This obsession stems from an adolescent trauma. The only son of rich, self-absorbed parents, he was once picked up by a homosexual chauffeur, Lino (Pierre Clementi), and taken to the invert's garret flat. Prior to seducing the lad, Lino let down his hair (literally), dressed in a kimono and invited Clerici to 'shoot the butterfly'. Clerici, scared, pulled the trigger, and ran away believing he had killed Lino. Guilt has paralysed him ever since, hence his anxious passion for passing as normal. This motivation is behind his marriage to a bourgeois girl, Giuglia (Stefania Sandrelli at her most coquettish), whom he does not love and attempts to betray, with Anna Quadri, following their arrival in Paris.

Moravia's novel, with its dense vein of introspection and its emphasis on destiny as a factor in the life of men, is not readily susceptible to screen transposition. Just how effective is Bertolucci's choice of cinematic equivalents then? For Moravia's sense of destiny, Bertolucci substitutes a psychoanalytic interpretation. Bertolucci adds a character who is important in his effect upon Clerici, a blind radio announcer named Italo, who introduces him to Fascist ideology and is finally denounced by Clerici, one the streets of Rome, on the night when Mussolini falls. Many of the novel's scenes depicting his lonely, deprived childhood Bertolucci omits, with the exception of the vitally relevant vignettes concerning Lino. Further elisions occur around the character of Giuglia. In the novel there is a wedding scene, significant because Clerici looks at Giuglia and suddenly feels a moment's tenderness – 'There had been a sufficiency of feeling and affection on his part ... of which he had thought himself incapable.' By omitting this scene, Bertolucci chooses to diminish Giuglia's importance. The possibility that a bourgeois marriage could offer any satisfactions is not something Bertolucci can yet contemplate, however glancingly.

Most sweeping of the differences between book and film is the ending. In the film, Clerici denounces his former friend Italo and turns his attentions to the eerily bleached creature squatting in an alley, who is none other than Lino, the nightmare figure from his adolescence. So Lino is not dead after all? All these years of guilty sterile living have been unnecessary. In the book, Clerici runs into Lino in a park, where he is celebrating Il Duce's downfall by making love in the bushes to Giuglia. 'When I met you I was innocent,' he tells Lino. 'And since then I haven't been.' Lino replies, 'We all lose our innocence. It is normal.' The film ends, however, as Clerici ogles the youth whom Lino had been soliciting, and blinks, as though he has suddenly come to his senses, and accepted his own homo-sexuality.

The Conformist seems a more robust film than *The Spider's Stratagem* not only because it breaks free of the Parma dependency – that nostalgic reversion to the childhood locale to which, even today, Bertolucci seems attached, but also because it opens new areas of emotional expression in the film-maker. Female sexuality is vitally celebrated in the movie – one of the most memorable scenes is the conga danced by Anna and Giuglia around the café while Quadri and Clerici, the quintessential non-dancing males, sit anchored to their table. Fascism, then, stems from an imbalance in the mind. It is a form of sexual aberration first, and a perversion of civic and bureaucratic duties after. This theme has been forcibly rendered by others in Italian cinema, most notably of course by Pasolini in his final movie, *Salò*. Bolognini has also tackled this idea, however, and the affinities of purpose between *The Conformist* and Bolognini's *Per le Antiche Scale* (*Down*

Stefania Sandrelli and Dominique Sanda performing the most vivid dance in Bertolucci's repertoire before a baffled crowd of Paris intellectuals – *The Conformist*

95

the Ancient Stairs), made in 1976, are worth consideration.

Down the Ancient Stairs, adapted from a story by Mario Tobino, is set in a mental hospital near Florence in the Twenties and takes as its theme 'asylums not as institutions where the mentally insane are cared for, but where the growth and acceleration of the processes of mental disintegration take place', in the words of the Bologini anthology published by the Italian Ministry of Foreign Affairs. The protagonists are Dr Bonaccorsi (Marcello Mastroianni), who has run the asylum for eight years, and a novice doctor named Anna, who arrives with quite different ideas from his. Anna believes in the caring therapeutic community. She sees that Bonaccorsi manipulates not only the patients but also the women, who are either doctors or doctors' wives, in order to disguise his own incipient dementia. Three women are his mistresses: Francesca, the mournful wife of a fellow doctor, who comforts Bonaccorsi during thunder storms; Carla, a doctor who plays sly erotic games with him but heads into town for her real sexual kicks; and Bianca, his devoted assistant. Anna has no intention of joining Bonaccorsi's harem. 'You want to make others believe that you're concerned with them because you're obsessed with going mad,' she tells him bluntly. 'You are incurably ill.' Bonaccorsi retaliates by telling her she is sexually repressed. Anna is strong enough to admit it but adds that she is not as 'open' as Bonaccorsi; she implies that she is not desperate enough to seek a solution in promiscuity. She clashes with Bonaccorsi on a theoretical level too. Insofar as his ideas have any coherent pattern, they seem behaviourist; he seeks 'a black spot' in the blood samples of his schizophrenic patients, believes eventually that he has proved his theory, only to have Anna demolish it – he simply did not clean his test-tubes.

This aspect of the film is somewhat gimcrack, as is the catatonic sister of Bonaccorsi's, who lies in a cell and receives scant medical attention. These *grand guignol* diversions aside, much else in the film is capably controlled, while the opening sequence is masterly, showing a patients' ball, presided over by Bonaccorsi. With the elaborate tracking shots, the music, the masks, the shifting colours and groupings, the scene is like a Visconti pageant gone awry. 'The world doesn't exist,' proclaims one patient, known as 'The Federal'. 'It has never existed ... Even I don't exist. It's as if I was dead.'

Bonaccorsi's regime, like Mussolini's outside, has led to acute dissociation, in which the will to personal dignity can appear only in wistful fragments, as with the patient who is writing a novel about Waterloo, but wants Napoleon to win. Bolognini's masked ball reveals the great pathos of sick people trying to play social roles. Clowns cavort near banks of flowers. The doctors sit separate at a high table. The music stops for medications to be issued. 'Among the spectators, mad or otherwise, there is great silence,' comments Aldo Tassone (*Image et Son*, No 217, May 1977). 'The film plunges our heads in our own inexistence ... Bolognini the secretive, the anti-exhibitionist, the realist, presents to us in symphonic manner the different characters and ... with great simplicity leads us to look at our own nothingness.'

At the end of the film Bonaccorsi, defeated by the common sense strength of Anna, concedes defeat and announces to his mistresses that he is leaving. 'I am frightened, I must be alone, I want to start again,' he announces. His defection is too much for Francesca. She leaves 'the circle of walls' for a final promenade round the town, and returns to kill herself by jumping from a high window. Her death reminds us of the danger inherent in Bonaccorsi's conduct; he has stimulated emotional dependencies which he cannot support. Anna's view of his inadequacy seems vindicated by Francesca's act.

Bonaccorsi leaves the hospital and goes to the railway station. He boards a train to the city with a bunch of *fascisti*, one of whom rants at him – 'We deny thought ... one thought for all.' Has Bonaccorsi understood that he has exchanged one form of madness for another? The camera closes on his enigmatic expression, as it does so often in Bologninian endings.

Yet life at the hospital reasserts itself anew without Bonaccorsi's disturbing presence. Anna walks in the grounds with the young patient, Tonio. 'You are happy,' she tells him. 'You just don't know it.' Another vulnerable young patient, Laura, who has a history of self-mutilation and anorexia, sits up and starts to eat. In these brief vignettes Bolognini suggests a subtle lifting away of psychic pressure. Recuperation, inside the circle of walls, might at last be possible.

Bolognini's vision of madness emerges, then, as different in emphasis from the depictions of such institutions under Fascism tendered by Bertolucci in *The Conformist* and Fellini in *Amarcord*. In atmosphere, the film is reminiscent of Rossen's *Lilith*, asking the same question, as Bolognini himself has put it, 'Why out of the blue does madness strike?' There is no answer to the question, but the willingness to see madness so directly is characteristic of Bolognini's engaging absence of doctrine or dogma. As Tassone rightly says, 'Bolognini's drama is non-extremist, in an Italy absurdly over-burdened with politics. They wish that he had Bellochio's rage, Fellini's surrealism, the austerity of construction of Visconti, the red-tinged intelligence of Bertolucci. But he is an original artist ... an intimate artist, far from fashions.'

During the 1960s Bolognini worked with another distinctive scriptwriter/ novelist, Goffredo Parisi, who adapted Moravia's novel *Agostino* for the screen in 1962, supplied the original story of Bolognini's arresting short, *The White Whale (La Balena Bianca)* in 1964, and wrote the screenplay for *L'Assoluto Naturale (The Natural Absolute)* in 1969. Bolognini is a director who is willing to shape his visual gifts around the concepts and themes of different writers. His films deriving from Parisi texts are thus very different in tone from those he evolved at the beginning of the Sixties, say, with Pasolini. If this suggests a chameleon quality to Bolognini, it should be added that he has always insisted on adapting texts to which he is strongly drawn, whether they be well-known Italian novels or original screen stories. Bolognini's relationship with Italian literature is flexible, but personal.

The White Whale, his short which was released in Italy as one part of the portmanteau movie *La Donna E Una Cosa Meravigliosa (Woman is a Mar-*

vellous Thing), is a strange study of a love affair between two malign circus midgets, and their efforts to murder 'the white whale' – a large lady trapeze artist to whom the male midget, Eros, is married. Superficially, the film has a Fellinian ambiance, depicting behind-the-scenes big top routine as did *La Strada* and *Clowns*. Bolognini's study of circus passions is actually very different from Fellini's, however. There is no trace of Fellini's vaunted 'sawdust and tinsel' nostalgia. The film provoked a vehement reaction when it was first shown in Italy, at the Venice Festival. Bolognini remembers in his interview with Jean Gili how he and the midget stars of the movie had to flee from an audience incensed by what they saw as a callous exploitation of handicapped people. In fact, *The White Whale* is a film with a mordant allegorical base. It takes the *crime passionel* theme – that theme so recurrent in a number of films coming from cinema in the Mediterranean countries – and by reducing emotional involvement in the characters' web of passion and intrigue (a lustful, world-excluding relationship between midgets is charged with an ironic connotation after all) allows Bolognini and Parisi to stand at a distance from their subject matter. An obvious parallel can be seen with Herzog's *Even Dwarfs Started Small*. Both Herzog and Bolognini are clearly fascinated (and slightly appalled) by the sheer intensity of libidinal energy

Jean-Louis Trintignant in *The Conformist*

when concentrated into humans who are small but ferocious. Yet it also seems possible to consider *The White Whale* in relation to a *crime passionel* like Visconti's *Ossessione*, in which 'big' lovers allow sexual passion to obliterate their moral discrimination. There is, in fact, a curious parallel between Bolognini's often operatic dramatic sense and that of Visconti. Both, at intervals, have shown a fascination with the somewhat overblown devices of opera and melodrama; yet where Visconti confronts his dramatic material directly, Bolognini favours an obliquer, more ironic stance.

The Natural Absolute (also known in the UK as *He and She*) is something of an anomaly, both within Bolognini's repertoire and the broader spectrum of themes generally made available by Italian cinema. It is a road movie, and that genre is as rare in Italy as it has been commonplace in the USA. It is, though, a road film of the most emblematic variety, concerning a struggle for sexual and intellectual domination between lovers who meet by chance, travel the *autostrada* together and then separate explosively, after the murder of one partner.

From the first sequence Bolognini signals the symbolical overtones of Parisi's screenplay. We see the anonymous hero, known as 'He', posing for his own camera amid Renaissance statuary in a public garden. 'He has doubt as his motto and oblivion as his reward,' drawls the hero. He is played by Laurence Harvey, who also produced the film. Briefly a woman ('She') is glimpsed across the garden. In the next scene he meets her in a car wash. She (Sylvia Koscina) is watching him from the wheel of her sports car. He approaches to ask, 'Why are you laughing?' She replies, 'Because you attract me.' They drive away together. She savours the moment when his camera snapped her in the garden – 'click click, like a mating call'. They make love in a motel. Immediately afterwards, the struggle becomes evident. He is in love with his imaginings about her, the eternal woman. His passion is secondary to the satisfaction he derives from feeding off it. She finds this infuriating. 'I'm a real woman,' she insists, back on the road. 'You can't invent me.' She is less a 'real woman', however, than a paradigm of Parisi's representing the Predatory Female. 'Why should I share my bed with thousands of images?' she asks him. To which he replies, 'But a man expresses his life like this.' Most men do not, however. It is a weakness of Parisi's concept that he remains too close to his hero's self-satisfaction.

The Natural Absolute, symbolic *grand guignol* in the manner of Ferreri's *Seed of Man*, is terrain that stretches Bolognini beyond his natural competence. His formalism, in moments, is characteristically concise and elegant, notably in the avenues of statuary in the opening sequence, but the excessively cerebral approach to narrative and character does not suit him. For this reason the film can be seen as a rogue exercise which hardly demands detailed assessment. What is interesting, though, is the strong parallel of theme and intention which exists with Bertolucci's *Last Tango in Paris*, made three years later. Both films begin on the deracinated middle-aged male's stylized preparation for self-oblivion. Against Bolognini's mannered opening, which

Laurence Harvey in *A Natural Absolute*

poses Harvey on the sepulchre intoning his epitaph as the camera clicks, one can juxtapose Brando lurching into view shouting 'fucking traffic'. Both men are deteriorated, facing their lack of a future. The random meetings with the women follow, absurd, morbid events, for in these men the will to find a mate has lost its enchantment, and the rapid couplings which follow in anonymous rooms are an expression of this ennui.

Bertolucci has remarked that he sees *Last Tango in Paris* as a political film because 'it's about a facing up, a relation of force between a person without a past and a person made up of nothing else.' But the film is less metaphorical than this makes it sound. The politics of *Last Tango* can be seen in two ways: there is the crass iconoclasm of Paul, a man who has never been able to relate himself effectively to the social institutions, as his numerous diatribes about his past life make apparent, but who has also failed to develop a strong sense of his own individuality, that vital defence the alienated non-citizen needs to call upon if social isolation is not to wreck him. There are also the socially conditioned assumptions of Jeanne to take account of. Jeanne is a girl from the upper middle-class administrative sector. Her father was a high-ranking Army officer in Algeria. Her widowed mother lives in a villa, with servants, outside Paris. Her boyfriend belongs to the radical chic intelligentsia; he is making a *cinéma verité* movie about her called 'Portrait of a Girl'. From the film-making fragments we see, his uncritical enthusiasm for the film-making process implies a movie project as rhetorically naïve as Bertolucci's *Partner*.

There is a deftly realized scene in *Last Tango* in which Jeanne, attempting to dissociate herself from the turgid erotic relationship with Paul, goes back to her mother's house to 'borrow' a revolver which belonged to her father. Equally telling, however, is her reaction to the urchins whom she catches trespassing in her mother's garden.

Her sense of property asserts itself unconsciously in this flurried act, but she lets them go as soon as the maid appears to add weight to Jeanne's rebuke. Jeanne switches from the authoritarian to the anti-authoritarian roles unwittingly. This small act is amplified, of course, in her flitting between her two roles with the men in her life, Paul and Tom (Jean-Pierre Léaud). With Paul, she plays the novice at the altar of Priapus – or 'happenis' as Paul would prefer to call it; with Tom, she plays the novice in his equally arcane rite – the film-making process.

Laurence Harvey and Sylva Koscina in *A Natural Absolute*

This 'doubling' of role is central to the psychological structure of *Last Tango*. It is a recurrent motif in Bertolucci's cinema and can be traced back to *Partner* and *The Spider's Stratagem*. Paul, too, shows similar symptoms of social schizophrenia. This is most marked in his encounter with his wife's former lover (played by Massimo Girotti), with whom he fences verbally in one of the flophouse interludes which punctuate his sexual liaison with Jeanne. Bertolucci underlines the point by having both men wear the same dressing-gowns. As Paul leaves, he says, 'I don't know what she saw in you.' The allure of the lover is obvious, however. He is a man who never pushes himself beyond his limits, as Paul does.

Last Tango provoked elaborate rhetoric from critics but little analysis. Pauline Kael, writing in *New Yorker* in 1972, described it as 'the most power-fully erotic movie ever made, and it may turn out to be the most liberating'. Norman Mailer, in his essay on the film, *Transit to Narcissus*, suggested that 'It will open an abyss for Bertolucci ... if he is going to fuck God, let him really give the fuck.' Mailer's doubt concerned the constraint Bertolucci imposed upon the improvisations – 'if the actors feel nothing for each other sexually ... then no exciting improvisation is possible on sexual lines.' This observation seems more valid for Schneider than for Brando. For Paul, laconic sex seems plausible, given his desperation; for Jeanne, the film's absence of physical as well as emotional tenderness during the act tends to shrivel her femininity. The erotic effects, as in Bolognini's *The Natural Absolute*, are part of the mannerism of the film. Both directors strain for painterly compositions during lovemaking scenes, bathing their respective 'boudoirs' in 'uterine' red light. The films' style, lacking any resplendent emotions to elaborate, withers into mannerism. Even the lavish accoutrements of Bertolucci's first truly internationally orientated movie – Bacon's paintings, Gato Barbieri's soundtrack – tend to underline the sterility of the characters.

The dialogue, too, contains its self-seeking element. This is most noticeable in the glibness of some of Jeanne's speeches, as in, 'I'm in love. He is full of mysteries. He is like everybody, but at the same time he is different ... he frightens me ... because he knows how to make me fall in love with him ... He's you. You're that man.' To which Paul responds by telling Jeanne to trim her fingernails, and launches into his famous sodomy monologue. *Last Tango* seeks to make the point that a girl like Jeanne, over-susceptible to men who believe themselves to be warriors and heroes, is bound to be manipulated by them. It is, after all, not only Paul who tries to invade Jeanne's soul but also her boyfriend, Tom, who betrays her affections, as she sees it, when he chooses to make a film about their affection rather than live it through. In this light, the ending of *Tango* becomes significant not just for Paul's dying gesture, parking his chewing gum as he dies on the high balcony overlooking Paris, but for Jeanne's broken mumble as she holds the gun ... 'I never knew him. He tried to rape me. I never knew him.'

The irony of the film, given our eight year retrospective view on it, is that

Marlon Brando in *Last Tango in Paris*

Bertolucci/Arcalli undercut their basic point about the imperialism of man towards young women by failing to grasp that they, too, in their screenplay, patronize Jeanne, giving her the improbable lines and excessively devotional emotions. Yet the film itself exaggerates the flaws in the character concept of the script. The secondary liaison, between Jeanne and Tom, contains the clues to Jeanne's own sexual aggression and confusion, and in the script this element is thoughtfully, if somewhat schematically tailored to counterpoint the apparent sado-masochism of the Paul/Jeanne tango in the Plessy apartment. In performance, however, Brando unbalances the point-making functions of the script. One is left with a sneaking suspicion that Bertolucci's coherence as director was over-taxed, that he was out-manoeuvred by Brando's solipsistic insistence that Paul was a tragic hero rather than a morose and loutish loser.

Bertolucci's gifts are real, but frail; his lyricism bruises easily and discolours. After this, his second movie shot in Paris, he must have felt the need to recuperate. He turned yet again to his home region, Parma, for *1900*.

Bertolucci took two and a half years to prepare the script of *1900*, in collaboration with Franco Arcalli and his brother Giuseppe. The film was eleven months in production and cost over five million dollars. 'It is a monument to the contradictions of the system,' Bertolucci stated, somewhat glibly. 'Between the capital which supports the movies and the idealistic themes of many talented directors I tried to enter and make an explosion.' The 'explosion' of *1900* is by now familiar: Paramount's insistence on over two hours of cuts, Bertolucci's refusal, lawsuits arising with producer Alberto Grimaldi, and then the 'historic compromise' in 1977 – a version running at four hours twenty minutes which was acceptable to both director and producer.

Bolognini's political epic, *Metello* (1970, based on Vasco Pratolino's novel of 1954), while it shares many affinities of theme and intention with *1900*, was produced under very different conditions. Bolognini commented on the stringent production measures to Jean Gili: 'I could only finally realize the film by accepting extreme economy of means ... the whole film was made using friends.' While the financial austerity which strictured Bolognini's epic

Gerard Depardieu as Olmo Dalco, socialist and guerrilla fighter, seen here exhorting the peasants, in a scene from *1900*

shows little in the completed movie, it helps to explain the limited European distribution of the movie, which has still not been shown commercially in London and Paris.

Both *1900* and *Metello* have in common, over and above their common political romanticism, a sympathetic stance regarding the late nineteenth/early twentieth-century socialist movement in Italy. For Bertolucci, the significance of his film lies in the present relevance of its political message – 'To speak of Fascism is to speak of the present,' he has stated. For Bolognini, *Metello* is a synthesis of his feelings about the relationship between political imperatives and human sentiment. 'I was much criticized because the film was sentimental and not political enough,' he has commented ... 'I'm accused of moving people when I ought to pose political problems. But I'm not a politician. The story I tell interests me as a human condition, then a political reflection ... I don't make films as a man of politics but as a cineaste.'

Bertolucci was also criticized in similar terms. A typical abreaction from an elderly communist official, Giancarlo Pajetta, was that the film 'glorified' the days of liberation in the spring of 1945. Bertolucci comments, 'the dance, the flags, the music. All this made him embarrassed. He said "We never did this, there was no explosion of joy."' Bertolucci adds, 'I tried to tell him that this day was not 25th April 1945 but a day in the future.'

1900 is the story 'of two destinies, of boys born the same day in 1900 who traverse the twentieth century right up to our time', as Bertolucci has described it. In fact, perhaps due to the excisions, there is nothing in the film of 'our own time' except the final scene, which shows the two boys as aged, amiable codgers. *Metello*'s span is somewhat earlier, say the 1880s to the 1900s, thus some but not all the historical referencing overlaps. The two boys of Bertolucci's story are Alfredo Berlinghieri (Robert De Niro), the son of a wealthy farmer in Parma, and Olmo Dalco (Gerard Depardieu), son of a peasant who grows up on the Berlinghieri estate, becomes involved in socialist agitation, leaves the area and then returns to lead the peasants' tribunal which tries Alfredo at the time of Liberation in 1945.

The resonances with *Metello* extend beyond theme to character, for Metello Salani (Massimo Ranieri) is also a working class child who grows up fatherless from an early age. His father is an anarchist agitator in Florence (where the story is entirely set) named Poldo. In the first scene we see Poldo released from prison, where he has been detained for political offences, to be met by his resigned wife and the baby son he has not seen, Metello. Poldo's wife dies a rapid death, simply exhausted, and Poldo is drowned in the river Arno five years later, while his son watches from the bank. As in *1900*, the narrative proceeds with rapid time lapses. We next see Metello, now a young man, refusing to emigrate to Belgium with the relatives who have brought him up. 'Emigration is death,' he says, backing away from the train, returning to Florence. He takes a job in the street market working for a socialist porter who knew his father, moves on to better paid work on a building site, and studies in his free time, already politically committed,

Robert De Niro and Donald Sutherland in *1900*

believing in the inevitability of the coming revolution. A veteran anarchist friend, however, isolates the current anxiety felt by many of his political persuasion – 'I'm no longer in agreement with anyone. They're all turning to growing onions and tending a flock of sheep. That's not what I fought for.'

The veteran is arrested and, when Metello tries to visit him in gaol, he is also thrown inside – guilt by association. In his large communal cell Metello learns from a social agitator that the anarchists belong to the past; now socialism is the battle of the future. *Metello* is deft at sketching this interesting shift from individualist to collectivist radical politics – the shift which, in British terms, was marked in that era by the fading of William Morris's political influence and the growth of the Labour movement.

Released from prison, with a record, Metello has a brief affair with a liberated widow ('Have you got a girlfriend? Pity. I'd like to have stolen you from somebody.') He is conscripted into the Army – '1,095 days without thinking,' he writes to Viola, the widow. 'Just cartridge cases to polish and an oath to memorize' – and returns to his former work, bricklaying, at a time of increased militancy from the workers. During a site rebellion one worker falls from a scaffold and is mortally injured. He tells the boss, Badolati, 'I'm not mad at you. You're the boss. You lost your leg too.' But to Metello he whispers, dying, 'Bury me under our flag without priests.' This separation of mere anger from a matter of socialist principle is his last act of discrimination, a far cry from the sweeping anti-fascism of Bertolucci's peasants in *1900*. The scenes which follow, however, might be the visual prototype of *1900*'s street violence. Socialist mourners wave the red flag by the coffin. Mounted police intervene. Collective action becomes a street brawl. Metello and several others receive prison sentences of eighteen months, but Metello has fallen for the daughter of the dead worker, Ursilia (Ottavia Piccolo), and later they are married.

This socialist romance, less truncated than that which occurs between Olmo and Anita in *1900*, is nevertheless similar in inspiration. Lovers who share the common cause are less prone to anomie than couples in a bourgeois marriage, Bertolucci and Bolognini mean to imply in their portrayal of these liaisons. Anita, the socialist schoolteacher in *1900*, organizes a literacy scheme for old peasants in a hostel which is incinerated by Attilio's Fascists. Ursilia, during Metello's periods in prison, takes over the wage-earning and child-raising, as she does during the long workers' strike which occurs towards the end of the drama. It is interesting that both directors want to show the reactions of these women when infidelity, on the part of Olmo and Metello, threatens their tenuous security. Both women are shown to be capable of forgiving; despite hard conditions, their dignity is ample.

The building workers' strike which comprises the final scenes of *Metello* is handled with considerable realism. Absence of strike funds is shown to be a major problem. Equally critical is the situation which arises when a relative of Metello's, who has returned from Belgium and started work on the building site, threatens to join the blackleg force which is due to move in shortly. As long as the repatriated worker has no wage, he argues, there is no prospect of bringing his family back to Florence. He is weakened by his deracinated situation, as Metello is weakened by his affair with Ida, the coquettish neighbour. Bolognini stresses, via these scenes, the importance of presenting no vulnerable areas in crises of this order, in which the vested interests are defied. By comparison with the operatic utopianism of *1900*, such a message may seem mundane, but is surely valid. Metello, arriving late for a critical strike meeting, fails to stem his relative's sudden panicky defection to the blackleg contingent. Next day, when the pickets and blacklegs fight by the gates of the site, Metello's close friend Renzoni is mortally wounded. 'It's not right,' he says. 'Getting killed for seven cents.'

The narrative might have ended here. Instead, we see Metello once again imprisoned, Ursilia struggling on her own, and finally going to meet him, new-born child in her arms, as he is released from prison. The child, a boy, is named Libero. Thus Bolognini ends on his favoured cyclical device, Metello's jubilant release echoing the opening sequence of the movie.

From this résumé, it is clear that there are distinct parallels between

Metello's political development and that of Olmo Dalco in *1900*. Both grow up in pre-militant circumstances, within an extended family, one proletarian, the other agrarian. Both learn to labour, develop political awareness, marry staunchly socialist girls, learn about sexual betrayal the hard way, and risk their lives, and the well-being of their families, in their fight for better conditions. Both arrive at a point of resolution, in which the political struggle they have participated in appears to reach a plateau, with the workers' claims to some extent vindicated. Metello helps the strikers win higher wages; Olmo takes part in the euphoria of Italy's liberation from the forces of Fascism.

Where the two political epics differ most markedly is in the *auteurs'* treatment of the vested interest groups who form the opposition. Bertolucci,

Stefania Sandrelli as the socialist schoolteacher Anita teaching the old men to read and write – *1900*

in part two of *1900*, dramatizes the rise of Fascist reaction to increased militancy in the blackest terms. Attilio, the Fascist foreman to the present ruling Berlinghieri, is a thug of the worst order, prone to murdering widows, cats and children. Bolognini's bosses, particularly the builder Badolati, are shown to be self-interested and grasping but not yet brutalized. The bosses of *Metello* are paternalistic in their repression of working interests rather than histrionically despotic. As Gavin Millar pointed out, à propos *1900*, 'It deliberately exaggerates and caricatures the horror of Fascism and, by the same token, elaborates its hopes for throwing off the Fascist and capitalist yoke with a visionary gesture of a brave new world which virtually tips into fantasy' (*The Listener*, 30th March 1978).

The political statement of *1900* is in fact more wily than most critics have allowed for. 'We are all born equal,' shouts Alfredo Berlinghieri Senior to the peasant patriarch Leo Dalco, pushing across the champagne so that they may both celebrate the births of their respective grandsons. Leo Dalco clearly does not agree but he takes a drink anyway and passes the empty bottle ironically to Rigoletto, Alfredo's hunchbacked buffoon, raising a laugh among the watching farmworkers. Later, when the Berlinghieri family are beginning to feel the economic pinch, and are reduced to toiling in the fields themselves, watched by malicious striking peasants, old Leo Dalco can hardly believe his eyes. 'Is this socialism?' he marvels. 'It can't last.' The repletion of the moment wafts him into gentle death, under a tree, while his grandson Olmo hunts for leaves to fan him. 'I always loved the wind,' are Leo's last words, in a whisper.

Another scene in which Bertolucci's pastoral lyricism fuses beautifully with a human response to the possibility of social change brought about by collective action occurs during the funeral of the incinerated hostel residents (this scene closes part one of the epic). Anita, heavily pregnant, walks in the column of mourning workers with Olmo. The procession is vast. Anita looks on in awe. Like Leo, she has suddenly glimpsed in concrete terms the power of the people, if galvanized. 'I'm not well,' she mutters, frightening Olmo, who thinks she refers to the baby. 'What a funny child you are,' she tells him, in the most perfectly rendered line in the movie – Stefania Sandrelli in her most empathetic relationship to date towards a Bertolucci text.

Part two of *1900* leaves behind the pastoral lyricism of Olmo and Alfredo's idylls in the Parman meadows – 'socialism with holes in the pocket' as Olmo, a beady brat, calls their communal masturbation and girl-baiting exploits. The second part deals in the main with the rise of Fascism, and the involution of the Berlinghieri ménage, with Alfredo Junior now presiding over the billiard table while his chic wife lapses into alcholism and Olmo defects from the local political arena, having had one tussle too many with Attilio and his Fascist hoodlums. Ada, too, leaves the district, after Olmo's defection. Unlike Olmo, she never returns. Olmo's return is timed to symbolical perfection. He arrives on Liberation Day, 25th April 1945, and takes charge of the tribunal which is to try Alfredo for collaboration with the Fascists. 'I have one thing to say,' remarks Alfredo, gently, 'I never hurt anyone.' Since his assertion is reasonable, in its way (Alfredo's problem is passivity not callousness), Olmo is able to sway the verdict. 'He is the living proof that the *padrone* is dead,' he tells the tribunal. So Alfredo is acquitted, and the workers hand in their weapons and dance away under the red flag, leaving Olmo and Alfredo alone in the town square, watched by the child-partisan. Together they jostle, as they did when boys in the 'long summer', while the camera rises joyfully over the roofs to settle in the lane which leads to the railway tracks. Thirty years later, two old buffers are still jostling in the hedges. While Olmo listens at the telegraph pole (as he used to do as a child, listening for his missing father) Alfredo lies down on the tracks in front of an approaching train, as the boys did in that golden summer. This time, however, Alfredo lies across the tracks not between them. These closing shots, like those which end *The Spider's Stratagem*, suggest a time-warp. Not a day in the future so much as a day in the distant past, is the drift of these images of maturity and coalescence. Past, present and future as a seamless whole, this is the vision Bertolucci is attempting to convey in these mellow, almost Falstaffian images. There is a level of awareness at which political dissonance no longer matters, he seems to be suggesting.

Bolognini's *L'Eredita Ferramonti* (*The Ferramonti Legacy*) of 1976 was adapted from a story by G. C. Chelli by Bolognini himself, in collaboration with Ugo Pirro and Sergio Bazzini. Like *La Viaccia*, the drama hinges on a family quarrel over a disputed legacy. In this case the legacy in question is that of Gregorio Ferramonti (Anthony Quinn), a wealthy Roman baker, who has a daughter, Teta, and two sons, Mario and Pippo, and dislikes them all equally. This film restores to Bolognini the terrain where he walks most firmly; his moral shrewdness is never more effective than when he satirizes the rapacity of the suburban middle-classes. The profit motive, family plots, corruption of bureaucracy, and nepotistic business *combinazioni* make up the elements of *The Ferramonti Legacy*. Bolognini is free to interpret his text not in terms of political heroics but of connivance. All the characters act out of self-interest. At the centre of the drama stands the curious relationship which develops between Gregorio himself and the wife of his son Pippo, Irene (Dominique Sanda, giving one of her most compelling performances). Irene is a woman who wants power and knows that to achieve this she must first acquire money.

Gregorio, in his turn, is also powerless, though he has money and position. He has retired from business and his son and daughter have taken over, so he shares with Irene a sense of being removed from the thrum of commerce. The relationship is given piquancy, in Bolognini's lightly perceptive handling, by virtue of Irene's mixed feelings for the old man. She wants his money, but she also finds she rather likes him. By comparison with her appalling husband and even more appalling sister-in-law, Teta, he seems quite honorable. For Gregorio, affection for Irene becomes the obsession of his twilight years. He prepares a room for her in his flat which is touchingly attentive to his idea of what a young woman favours – lace doilies on the table and sunflowers in a vase are hardly Irene's style, but she comes to his bed anyway, albeit briefly.

The Ferramonti Legacy may seem like a far cry from the angry contempt for the owner class depicted in *La Viaccia* yet the distance is a signal of Bolognini's own development. The same themes of persiflage and calculation dominate, but now Bolognini's touch is lighter. Irene manoeuvres from patriarch to husband to lover (Mario, the other brother). Marrying Pippo, she steers him towards a portion of lucrative contracts being issued by his brother-in-law, a civil servant in charge of the tenders for repairs to the Tiber's banks. Soon Pippo prospers, but becomes no more gallant. Her affair with Mario is designed to increase the romantic aura of her new position as an affluent businessman's wife, but also to consolidate her own fortune, for Mario is an investment broker. 'I'll make you rich and respectable,' Mario promises her. But he, too, becomes boring, and so Irene switches her attentions to the patriarch, Gregorio. In the end, Mario, realizing that he has been betrayed by Irene, just as he himself has betrayed Pippo, shoots himself. Pippo declines into drunken paranoia. Gregorio dies of over-excitement in Irene's bed. All three Ferramonti males are vanquished.

Irene herself now comes to be crushed. The suicide of Mario leads to her appearance in court, to which she has been brought by Teta and her husband, who dispute the Ferramonti legacy being left to her. They win. 'Power always wins in the end,' is Teta's husband's jubilant conclusion. 'Let's drink a toast to progress.' Irene is left to face a future that looks, to say the least, equivocal. She had the will to succeed at anything but her intelligence was too narrow; she saw success only in terms of power and money.

With *La Luna*, in 1979, Bertolucci has gone full circle. If *1900* marked the outer limits of his expressive purpose, *La Luna* returns him to his core, the operatic theme which derives from the importance of family relations in Italian life. As usual in Bertolucci's cinema, the family is shown in subverted fashion; with this study of incest between mother and son he resumes the theme first established in *Before the Revolution*.

Two momentous crises of self occur in *La Luna*: the first, when Joey and Caterina, in their New York City apartment, look out onto the street and realize that Douglas's limousine slewed across the pavement is not a macabre joke, but a very ominous moment for them. Douglas, at the wheel, has died from a heart attack. Caterina has lost her manager/agent/husband; Joey has lost his best buddy/fellow baseball fan/father. The second occurs in Rome, where Joey and Caterina are now staying while she sings in the opera-house. It is Joey's birthday and Caterina is dancing at his terrace-garden party. She looks down to the street and sees Joey in an alley with his schoolfriend Arianna. Here she discovers Joey shooting up on heroin, with Arianna his accomplice. The look on Caterina's face sends Arianna quickly packing. Once again Caterina has been brought down from her cultural ivory tower by the brute facts of street life, as these now afflict her son. Husband dead, son an addict. To watch the way Bertolucci composes these two central crises of awareness is to watch a master craftsman.

This said, it remains to add that much of *La Luna*'s style is altogether less distinguished. In this film Bertolucci has apparently perfected his experiments in the use of what one might call the Gregarious Camera, which is in a state of constant motion around the 'obscure object of desire' – the actor. Storaro and Bertolucci have, over the years, honed the use of short thrusting tracks, pans and dolly shots into a very distinctive shooting style. This very restlessness of *mise en scène*, however, poses its own problems for Bertolucci. He is always on the outside of the crisis. A director like Bunuel sets up the camera and then films what has already been mapped out with a sardonic clarity in his mind, whereas Bertolucci's constant shifts suggest his own *irresolution* about what is happening to his characters.

This is not to deny that *La Luna* is, in parts, as good as anything Bertolucci has achieved to date. There are moments, odd lines and brief vignettes, which are intelligent and affecting. 'Are those ducks or geese?' cries a weeping Joey to his mother, in the hearse, during the funeral service. It's a marvellous line, delivered perfectly, although its clarity only serves to accentuate the overall inconclusiveness of the work.

Caterina, as focused by Bertolucci/Clayburgh, seems an oddly arrested personality, protecting herself from the world by means of an entourage – the assorted camp-followers and girlfriends who screen out tangible problems, not least of which is her failure to be a mother to Joey. There is not much difference between the modish narcissism of Caterina and the amorality of Gina, Fabrizio's equally disorientated aunt in *Before the Revolution*. Bertolucci, however, is eight movies on, and fifteen years older. Where, then, is the ripening of perspective one might expect from him?

Alongside Caterina's self-absorption, Bertolucci lays his parallel purpose, lending to *La Luna* obsessively placed echoes from his previous repertoire, and that of his own *maestri*. The influence of Pasolini, for example, is potent in the opening sequence. As in the opening scenes to *Oedipus Rex* (1967), Bertolucci's opening to *La Luna* reveals the primal situation, mother and infant son, the 'rejoicing in the sunshine', as Pasolini's screenplay puts it, 'a mysterious corner of the world where new life is stirring.' Both sequences contain the threat – the intrusion of the father figure who wants to take the mother away from the son, and over the love-making (which Pasolini chooses to depict and Bertolucci to elide) is the moon 'on the roads and peaceful fields'. This moon hangs above Caterina and Joey as they bike along the coast road, and then, in *both* films, comes the abrupt cut and time lapse to the child Oedipus and adolescent Joey being horse-traded among relatives who are hardly aware of them.

But if Bertolucci cannibalizes Pasolini's *Oedipus Rex* in *La Luna* he also cannibalizes his own movies. Echoes of each one are present. Joey's stepfather, Douglas, finding the wad of gum on the balcony, mutters, 'leaves his damned

Jill Clayburgh and Tomas Milian in *La Luna*

gum everywhere'. Who though? Joey, or the stickman of *Last Tango*? Having removed the gum (which Brando parked in 1972, on his balcony in Paris) Douglas himself dies in the next scene. The detail serves as an omen.

In the funeral scene which follows shortly, images from *The Conformist* are present, in those sinister faces crowding Caterina and Joey as they huddle in the limousine after the interment. Not sinister in themselves, these faces, for they wear dutifully solicitous expressions, but sinister by association, for the faces which similarly crowded Clerici and Anna Quadri as they huddled in limousines were those of assassins. Most conclusive of the many echoes is the finale to *La Luna*, in which Joey and his new-found father watch Caterina rehearsing Verdi's *Un Ballo in Maschera* at the Caracalla Baths, her voice soaring on stage while Arianna, Joey's schoolfriend, sidles towards him. *Before the Revolution* ended similarly, with Fabrizio at the Parma opera-house for Verdi's *Macbeth*; while his fiancée waits inside he says goodbye to his mother figure, the alluring aunt with whom he has had an incestuous affair. The same shift to normalization is implied by Bertolucci, or perhaps he is only shadow-boxing. With this scene, one senses that his cinema has become air-tight, a vacuum inside which icons are preserved. One might mention the almost fetishistic use of props from previous films, like the cow-shaped milkjug which Caterina wields at tea-time with Joey (last seen in Clementi's fist in *Partner*).

There is evidence that the Parman sequences of *La Luna* have been cut in the UK version; certainly the walkabouts around the opera-house, to which Bertolucci alludes in his interview with Richard Roud, are truncated. Cuts alone are not enough to explain the patchiness of the narrative in the latter part of *La Luna*, however. The turning point is surely the encounter between Caterina and Joey's dope dealer, the serenely malign Mustapha, in which Caterina secures her son's rations and learns a bit about him from Mustapha's judicious perspective. Mustapha, in his spotless white caftan, with his nude pin-ups on the wall of his spotless hovel, and his unctuous concern as he deals out the dope, represents the moment when Caterina knows she is deposed: heroin, now, is Joey's mother; and Mustapha its administrator. Mustapha's amoral courtesy is surely one of the script's more perceptive and unnerving moments? It suggests that the real evil of Eastern drugs in Western bodies lies with the false sense of self-esteem and power induced in those who traffic. When *La Luna* touches on the clash between such value systems it jerks to life. Perhaps Clare Peploe, Bertolucci's co-writer (and lately his wife), had a hand in these wry intervals.

Angela Carter's account of *La Luna*, the most strenuous examination to emerge from the British critics, states the feminist perspective on Caterina's derangement. When Caterina shows Joey where his father works, Ms Carter comments that this is 'the authentic irrationality of patriarchy, the triumph of nature over nurture, the consecration of seminal fluid as the supreme unction of human bonding'. Such elaborate sarcasm is wasted on *La Luna*. Caterina hands over Joey because she wants the boy's father-hunger off her conscience. In a sense, this is the rightful start, in dramatic terms, to the drama which matters – how Joey copes with having two parents. Unfortunately, Bertolucci chooses to end his film just when it could begin to command a denser level of attention. Bertolucci settles for a sequence which ducks the pertinent issues of the new element; a semblance, in Joey's life, of family.

We go to the Caracalla Baths with Joey. Caterina is rehearsing but can 'break' for five minutes. Mother sings again, father has been found; with Joey, he watches Caterina rehearsing on the wide stage above them. The camera soars again, but the reconciliation is intangible. Joey's avenue to self-determination, like Fabrizio's at the end of *Before the Revolution*, is only a cul de sac. Mother is still the Medea; even her voice belongs to Pasolini's Medea, Maria Callas.

In *Agostino*, 1962, Bolognini tackled the theme of potential incest between a beautiful young mother (Ingrid Thulin) and her teenaged son (Paolo Colombo) less frenetically than did Bertolucci in *La Luna*, but too openly for Italian taste in that era – the film was offered very limited distribution in Italy, although today Bolognini still regards it as one of his most perfectly realized films. The film was adapted by Goffredo Parisi and Bolognini from Alberto Moravia's novel of the same title, and shows the becalmed relationship between mother and son during a holiday in Venice. They boat and swim, living in a dream of mutual tenderness, until a man intervenes, and the mother responds, at first reticently, aware of Agostino's delicate feelings towards her, and then more overtly. Excluded, Agostino becomes involved with a gang of beach boys who teach him their way of life – dockside visits to whores, homosexual dalliance. This contact with a rowdy but life-affirming underworld enables Agostino to emerge from the dangerous cocoon of emotional tension he has woven around himself and his mother.

Bolognini catches with considerable moral and visual finesse these fervent, dreaming days in and around the Venice Lido. Drifting gondolas, derelict islands, a haunted cemetery and waterside cabins are his landmarks. The adaptation is faithful to Moravia's book and remarkably unfussy. Where the novel expatiates on the significant shifts of mood between Agostino and his mother, Bolognini/Parisi employ a laconic voice over technique. The characteristic Moravian blend of innocence and corruption is modestly conveyed, while the agitation in Agostino's inner life Bolognini represents by the use of startling dream sequences. While Joey and Caterina in *La Luna* indulge in the ritual antics of the 'New Dramaturgy' (less euphemistically, jaded Living Theatre), but never come to grips with themselves, Bolognini's study of an adolescent incest trauma sketches in the potentials for recuperation via the peer group.

Matthew Barry and Tomas Milian in *La Luna*

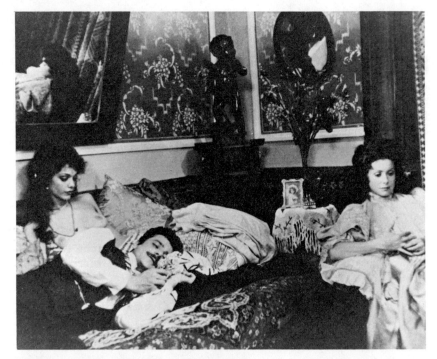

Tina Aumont, Giancarlo Giannini and Catherine Deneuve in *Drama of the Rich*

At the end of the Seventies, both directors appear to have turned away from their cinema of political rhetoric and returned to their essential concern for the more compacted turbulence of family conflict. Bolognini's *Fatti di Gente Perbene* (*Drama of the Rich*), 1979, re-enacts a notorious murder trial of 1902 which occurred in Bologna and concerned the well-to-do, free-thinking family of Murri. Tullio Murri (Giancarlo Giannini) is tried for the murder of his sister Linda's husband, Bonmartini, and sentenced to thirty years. This element, Tullio's emotional over-involvement with his beautiful, apparently oppressed sister, contains a latent incestuous motif, but the film's main velocity occurs on a more specifically political dimension. When Tullio is indicted, so are several others, among them Linda herself (whose ten year sentence is subsequently commuted by the King), Rosa Bonnetti, who has been Tullio's lover and accomplice in the preparation for the murder, and two other men, one a friend of Tullio's and the other a lover of Linda's.

Professor Murri, an academic scientist with influence over students at the university, is a subversive influence in the eyes of the investigating policeman, Stanzani, who makes it clear that his aim is not merely to identify a murderer but to call in to disrepute those liberal values the Murri household represents. Tullio himself is a force for potential change in

Bologna, a lawyer and prospective socialist candidate in the forthcoming elections. At the end of the film, when Tullio is sentenced to thirty years imprisonment and his accomplices to shorter terms, Professor Murri is devastated. Stanzani's investigation has amounted to a witch-hunt in which his own good name has been irreparably mutilated.

Drama of the Rich is thus a bilateral narrative, coupling Bolognini's perennial interest in emotional frailty with scriptwriter Bazzini's interest in the theme he extrapolates from court documents of the period – political vendetta directed by authoritarian forces against the Professor's freethinking philosophy.

Sergio Bazzini, incidentally, has worked with Bellochio, notably on *Victory March*. He also co-adapted *The Ferramonti Legacy* with Bolognini and Ugo Pirri, and Bolognini's respect for Bazzini's trenchant political beliefs as a scenarist serves, perhaps, to invigorate his vision now as his collaborations with Pasolini once did. Bolognini's strengths as a film-maker, today as in the past, reside in his compositional sensitivity, which has been helped by his regular work with two major cinematographers, Nanuzzi and Guarneri, and his quiet determination to make films which are rooted in humanistic and literary culture whenever possible, however difficult the production circumstances.

Bernardo Bertolucci during the shooting of *The Spider's Stratagem*

Bertolucci, now approaching his fortieth year, has made nine feature films to date and is able to command production resources outside Italy to an extent no other Italian film-maker, with the possible exception of Fellini, can do. For all that, his cinema, as it developed through the Seventies, has by no means fulfilled the great promise evident in *Before the Revolution*, *The Spider's Stratagem* and *The Conformist*. It is arguable that, with *The Conformist*, Bertolucci achieved a peak, in his taut control of every aspect of the film.

Federico Fellini shooting *Roma*

4. Fable and Phenomena: Pier Paolo Pasolini and Federico Fellini

In 1960 Fellini completed *La Dolce Vita* and established his international reputation while Pasolini began shooting his debut feature, *Accattone*. Both film-makers came from the North of Italy, both had arrived in Rome during the Forties, poor and hungry, living on the outskirts, and both had begun their careers in cinema working as scriptwriters, Fellini with Rossellini, Pasolini with Bolognini. It is their careers in the Sixties, and still more in the Seventies, which concerns us in this chapter, however, for during the Sixties the development of both *auteurs* underwent marked changes. The street-life exuberance of their early films harshened into a more complex vision, the forlorn waifs and rowdy *vitelloni* of Fellini's earlier films being remoulded into the emotionally complex figures of Guido and Giulietta in *8½* and *Juliet of the Spirits*, while Pasolini moved on from his 'boys of life' studies of apache-packs to the fiercer rites and rituals of the mythic world in *Oedipus Rex* and *Medea*.

The mystical waif who haunts Fellini's early work, most notably *La Strada* and *Nights of Cabiria*, is still present in the watershed film, *La Dolce Vita*, but is now a receding presence. At the end of the film, on the beach with the 'sweet life' crowd, the journalist-hero, Marcello, sees the angelic girl calling to him across a strip of water. He cannot hear; he is drawn back into the vicissitudes of the gang. In this, perhaps the bleakest of Fellinian endings, we see the hero emasculated and corroded by the corruptions of the Via Veneto party goers, whose outer gaiety masks inner chaos.

Beyond *La Dolce Vita*, in Fellini's two autobiographical studies, *8½* and *Juliet of the Spirits*, the transformation of Giulietta Masina from mystical waif to mystical matron is completed. Both films are formally fragmented but cohere as intense emotional statements, or visions. What imbues the two central relationships with such candour – those between Guido and his wife in *8½* and Giorgio and his wife in *Juliet* – is Fellini's intense nostalgia for an ideal of love as shared identity, what Montale, in his poem *Xenia*, referred to as follows –

> 'Yet it gives me no rest to know
> that alone or together
> We are one.'

At the heart of Fellini's nostalgia is just this sense of loss, and yet wonder. For what has passed away – the innocence of romantic love – remains as a residue of tenderness towards life. Guido is left, at the end of *8½*, in a circus ring made up of his own contacts – associates and artists in the cinema industry. He gathers the elements of his life around him. Juliet, on the other hand, is left entirely alone at the end, bereft of husband, friends and maid-servants. Yet she, too, is reconciled. That secret half-smile on Giulietta Masina's face reveals this. The difference in the two final scenes is perhaps Fellini's experience of the disparity between the life of middle-class Roman men and women. Or it might equally be Giulietta Masina's notion of this difference, for her presence in her husband's films has never before been so resolute.

Pasolini, too, shares this fundamental sense of poetry but stands in a more tormented relationship towards it. Where Fellini is complex. Pasolini has often appeared to be merely complicated. The writings, as well as the films, of the two men help to reveal their great affinity – their reverence for the myth and 'sacrality' of life (Bertolucci's term for his former mentor Pasolini's basic disposition) – as well as their considerable aesthetic differences. Fellini's 1967 essay on Rimini, written while he was recuperating from a major illness, is as evocative as Dylan Thomas' *Under Milk Wood*; both stand as classic portraits of a small town between the two world wars; while Pasolini's writing, whether the early novels such as *Boys of Life* or the later semiological essays about film form, teems and crackles with cerebral static.

What is most interesting about *Accattone*, Pasolini's debut feature, is the difference in emphasis it offered from the two 'boys of life' films Pasolini had scripted for Mauro Bolognini in the years preceding his own first feature. Pasolini's hero, Accattone, is more darkly shaded than the characters he devised for Bolognini in *La Notte Brava* and *La Giornata Balorda*. While both film-makers make similar use of the dusty *periferia* of Rome in summer, and both treat the pick-up as the central emotional intersection in their lives, Pasolini treats his apaches more emblematically. Accattone, who can go down no further, is a potential saint, given Pasolini's own values. He is too proud to take a job, too puny to triumph over those who despise him, and too tender-hearted to manage a frightened prostitute, whose earnings alone could save him from destitution. His moment of greatest shame occurs when he is obliged to steal his own son's necklace, in order to raise the money to kit out his apprentice whore in new clothing.

In the final sequence, Accattone is reduced to toiling round the inner city with two thieves. They filch some food from a truck. The police arrive. Accattone makes a dash on a stolen motorbike, crashes it (offscreen), and

Enrique Irazoqui as Jesus in *The Gospel According to Matthew*

is glimpsed as he lies dying. 'I'll be alright now,' he murmurs. Pasolini believes, then, that death is redemption. The ending is abrupt and potent.

Accattone is a pivotal movie, for it refers back to Pasolini's scripts for Bolognini, and forward to Pasolini's other films dealing with the male pack ethic, *La Ricotta* and *The Gospel according to Matthew*, the former portraying 'the boys of life' as film extras engaged in filming a crucifixion, one unfortunate being left hanging on his cross after shooting ends, while the latter portrays Jesus as an evangelizing leader of the pack, on the rampage through the hill-towns of Southern Italy, which Pasolini has said served him very well as a facsimile Palestine.

Paul Mayersberg, considering *Matthew* in *New Society* (8th June 1967), compared Jesus' rebelliousness with that of Accattone – neither worked, both had disciples. Mayersberg perhaps overstretched his parallel when he proposed that Jesus, like Accattone, was 'a procurer for God'. Pasolini himself contradicts this assertion when he remarks, to Oswald Stack, 'I am not interested in deconsecrating. This is a fashion I hate ... I want to reconsecrate as much as possible.' It now seems apparent that *Matthew*, too, is a pivotal film in Pasolini's development, easing him from his preoccupation with male pack values into a more aloof contemplation of worlds beyond time, myth-obsessed and pagan, in which the actions of the protagonists are determined not by Christian scruples but by the imperatives of rite and ritual – the polytheistic obeisances.

The late Sixties were in fact a period of creative turbulence for Pasolini, who counterpoised his adaptations of Euripides – *Oedipus Rex* and *Medea* – with his two highly personal studies of contemporary moral corruption, *Theorem* (1967) and *Pigsty* (1969). Fellini, as noted, during this period was also undergoing a change of focus. It now seems plausible to suggest that, without the fragmentation then occuring within the two film-makers as conveyed in these films, we might not have received the master works of the Seventies. For the shift from the fragmented, confessional cinema of the late Sixties to the more detached, satirical vision of the Seventies has led Fellini and Pasolini towards the rigorous challenge of shaping classic texts into cinematic formats, Fellini adapting Petronius' *Satyricon* and the *Memoirs* of Giacomo Casanova, Pasolini adapting in turn Boccaccio's *The Decameron*, Chaucer's *The Canterbury Tales*, *The Arabian Nights* (these three constituting his Trilogy of Life), and, finally, de Sade's *One Hundred and One Nights of Sodom*.

Reactions to the mid-Sixties shift from narrative to confessional-cinema were divided. Edward Murray, for example, who had responded warmly to Fellini's earlier work, such as *La Strada*, alludes scathingly in *Fellini the Artist* (1976) to *Satyricon*. 'The sequences could be reversed for the most part, without damage to the structure ... an aesthetic zero.' Against this view, we have the reaction of Jon Solomon in his book *The Ancient World in the Cinema* (Yoseloff/Tantivy 1978): 'Some pieces of the film may belong in a classroom, some ... in an asylum, some ... in the bathroom, but the entire creation is a brilliantly individual impression of the spirit of the ancient

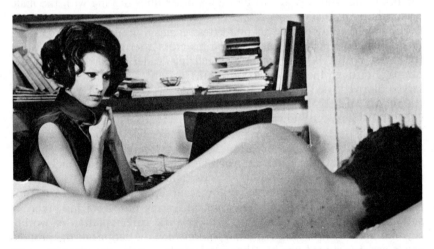

Silvana Mangano as the rich housewife beguiled by the mysterious interloper (Terence Stamp) in *Theorem*

Pierre Clementi (Foreground) in *Pigsty*

novel.' A view, by the way, with which Paul Barker concurred, reviewing the film in *New Society* (17th September 1970) – 'It is the best replica of the feel of Petronius, and the rest, that I have come across.' Between these poles of opinion, the appalled and the awestruck, *Fellini's Satyricon* has had to settle.

Fellini himself, considering the film in the preface to the screenplay (published in the UK in *Fellini, Essays and Criticism*, OUP 1978), described the revolt of his young heroes Encolpio, Aeschyltus and Gitone as an aspect of 'their utter ignorance and estrangement from the society surrounding them ... they eat, make love, stick together, and bed down anywhere ... they are totally insensible to conventional ties like family ... they don't even practise the cult of friendship, which they consider a precarious and contradictory sentiment, and so are willing to betray or dismiss each other at any time.' Such acts of betrayal and dismissal are of course part of the humour of the movie, notably in the cave of the minotaur. The comic impetus of *Satyricon* derives from the cruelty of the protagonists, these pre-Christian *vitelloni* who contain an incorrigible mixture of vitality and depravity.

Edward Murray's criticism that *Satyricon* lacked a linear narrative seems off the point, since the value of the film, for Fellini, as for his audience, is that it is an experiment in cinema as 'unfinished fresco', as he has described it. We might recall, too, that the Petronius text was itself only a fragment of a much larger (uncompleted) novel. Although there is no recurrent plot mechanism, there is considerable dramatic tension in the picaresque exploits

An orgy scene from *Fellini Satyricon*

of the youths as they journey from the bowels of Rome to the open beaches facing Africa.

What *Satyricon* lacks, in fact, is not so much suspense (there is plenty) but a clearly discernable morality. The plot framework, which usually serves to activate the author's attitudes towards his characters, is missing. The only scene in the film which contains what one could call 'conventional' (i.e. – identifiable) sentiment is the episode at the villa of suicides in which a wealthy couple send away their children and commit suicide before the barbarians arrive to pillage. There is a dignity, both visually and aurally (in Rota's otherwise magnificently ferocious score), about this scene which occurs nowhere else in the film. Does Fellini, then, intend to suggest here a foreign 'Christian' feeling of self-abnegation which puts to shame the licentiousness of the rest? Or does he intend us to view the scene satirically, as a parody of every middle-class couple's self-sacrificing grievance, 'we did it for the children'?

The amorality of Fellini's characters is nothing new. As long ago as 1963 Gilbert Salachas observed, 'Fellini never becomes a prosecutor who condemns his characters ... nor does he ever serve as a devoted lawyer who gives them unconditional support ... it is an ambiguous position, one which shocks people who like things clearcut.' Exactly. It remains to add, however, that it is Fellini's freedom from cant that underwrites his continuing vitality as a major film-maker, no less so in the Eighties than in the early Sixties.

There is a kind of 'conclusion' built into the *form* of the film, however, as is so often the case with Fellini's later films. Each 'road to freedom' explored by the heroes turns out to be a cul de sac – Vernachio's theatre, Emolpio's salon, the minotaur's cave, the cavern of Oenoethea. So our heroes are left on the beach, one dying, the other about to leave on a boat for Africa – another potential 'road to freedom' which is likely to prove disappointing. They have superseded those mournful men and women who were once Fellini's protagonists. Zampano, Marcello, Guido and Augusto have been somehow gusted away by Fellini's new mood of nihilistic laughter. *Satyricon* leaves Fellini clear to 'do what he likes', as one admirer of the film, Henry Miller, put it in a letter to Fellini.

Pasolini's *The Decameron*, made the same year as *Satyricon*, explores in comparable manner the idea of film as fresco rather than visual text. Pasolini makes use of the device of the travelling mural painter, a pupil of Giotto's. As the film ends on the vast mural the artist has completed in a cathedral, he mutters wistfully, 'why bother to complete a work of art when one can dream it so much better?' Like Fellini, Pasolini ends on a frieze; unlike him, he feels the need to comment (he also plays the painter). This is the key to the affinity, and the difference between them: both are drawn to what Pasolini calls 'kinemes' – images which are purely the province of the cinema, deriving from dreams and daydreams as much as mundane reality, yet Pasolini has always been more attached to analysing the transposition of dream to cinema.

During the filming of *The Canterbury Tales* in England Pasolini said, 'The works are very merry. They all look at sex without inhibition, with freedom, and above all they're enormous portraits of the origins of entire nations.' He also commented, in *Image et Son* (No 267, January 1973), that with the trilogy he had returned 'to my first film, which was born instinctively. I've renounced an explicit ideology and tried to make something which would be only cinema.' We should not take Pasolini's description of the film too literally, however. The violence, connivance and duplicity depicted in the stories suggested a blacker vision than the one Pasolini describes. Among the atrocities contained in all three films are sexual perversion, the incineration of homosexuals, poisoning, vendetta murders and castration. The critic who came closest to discerning the essential Pasolinian misanthropy in the trilogy is Gavin Millar, who noted, in *The Listener* (2nd March 1977), 'His anti-clericalism and his general mistrust of male-female liaisons is the key to Pasolini's stance. Men are brutal, casual and unfeeling in their dealings with women; women are deceitful and self-deceiving in their relations with men. Pasolini perhaps derives less fun from Boccaccio than from the revelation of hypocrisy.' One can in fact argue that, contrary to Pasolini's assertion that the trilogy satisfied a hunger in him for a less bureaucratic world, the films are not nostalgic at all, but satiric. Using the medieval theme as his vehicle, Pasolini again tries to ram the fortress of late twentieth century life, with a wheelbarrow not a tractor.

Jenny Runacre in *The Canterbury Tales*

Pasolini's force as a critic of his country's institutions stems partly from his homosexuality and partly from his eccentric political ideas, which corresponded less to the orthodox Left than his own brand of 'right-wing anarchism' (as Norman Mailer has described his own volatile political disaffiliations). In the late Sixties, for instance, Pasolini wrote a poem in defence of a policeman, claiming he preferred the iniquities of the police to the hypocrisies of the student radicals. From his marginal position in society came his exuberance, but also his narrow range. Towards the end of his life his published writings, even more than his films, reveal a tone not dissimilar to that of other great cultural pariahs such as Neitzsche and Wittgenstein. 'Back to the rough ground,' said Wittgenstein. 'What we are destroying is nothing but houses of cards and we are clearing up the language on which they stand.' This all-embracing iconoclasm Pasolini would have understood, and approved of. To him language, whether written or cinematic, was as tangible a reality as family is to most people. From this stems the curious impression his cinema makes – Pasolini is peerless but somehow remains emotionally thwarted.

The scepticism of Boccaccio and Chaucer Pasolini also shares: we see priests sanctifying killers on their deathbed, sodomizing one another and burning their brothers, while nuns frolic lecherously with youths in convent gardens. That the Church of the era was involved in venery is established by Andrew McCall in his book *The Medieval Underworld* (Hamish Hamilton, 1979), in which he points out that the Papacy leased premises for brothels. Venery outside the churches, in the villages, is established by Emmanuel le Roy Ladurie, in his research notes entitled, *Montaillou, village occitan 1294– 1324* (Gallimard, Paris 1976), in which he proposes, with regard to the village of Montaillou, 'We are faced with a complex web of relationships within an extra small community [of fifty married couples]. Perhaps it is natural we should find many couples engaged in sexual relationships outside marriage, and a frequent preoccupation with incest.'

The rudimentary nature of Italian capitalism at this time must also have appealed to Pasolini. In the letters of a fourteenth-century merchant named Dantini, we learn that when a bank was opened in Florence in the 1370s Dantini was arrested on charges of usury. At that time, he records, taking interest on any loan was forbidden. In addition, usurers were excluded from the sacraments – there was a riot after one usurer was buried in a chapel. This popular mistrust of usury – that component of capitalism now so integral to its *modus operandi* – again helps to explain Pasolini's uncanny empathy for the era.

This empathy is not only intellectual, however, but also extends to Pasolini's formal control of his medium. 'He comes at The Tales at an angle without explanation,' remarked Gavin Millar, 'so that we are thrown into a situation which is getting on with itself … without paying us much attention.' We can see the same facility of presentation in earlier Pasolini movies; in the opening shot of *Oedipus Rex*, for example, Ruzzolini's camera pans off a signboard saying Thebes with all the snap of a Raoul Walsh opening. Again, in *Medea*, in the opening scene in which Jason is instructed on the power of destiny by the centaur, as they sit by the river, we see Pasolini treating this mythic transfer of hidden knowledge quite naturalistically, as though myth were his domain, and the modern corporate world alien ground. Pasolini engages in this fabled worlds at street level, so to speak, and in this respect differs from Fellini, whose treatment of the ancient world, or the Enlightenment, is stylistically less accessible. What both *auteurs* share however, and in this respect Franco Zeffirelli also possesses great ingenuity, is the eye for the theatricality of the human face. In *The Canterbury Tales*, Pasolini's use of British faces (those of well-known actors and stray hitchhikers he picked up *en route* across England) comprises a remarkable gallery of rogues, unlike any made available to us by a British film-maker.

Franco Citti in *Oedipus Rex*

Finally, on the theme of Pasolini's confidence with his texts, we should mention his capacity to add his own flourishes to the adaptations. Ciappelletto, for example, the menacing rent collector of *The Canterbury Tales* (played by Franco Citti), is a Pasolini invention; likewise, in *The Decameron*, Pasolini inserts the figure of Giotto's assistant (played by himself). This character provides the concise parallel with our times, for the Renaissance painter, with his teams of colourgrinders and mixers, was very like the filmmaker, heading a small, tightly-organized team of experts in order to produce works of art on a collaborative basis.

Of the three films, *The Decameron* is, in visual terms, the most decorative and lyrical in its use of colour, landscape and music (it is also the one film of the three shot in Italy, mainly around Naples). But even *The Decameron* has an ambiance far removed from the guild medievalism of, say, William Morris or D. G. Rossetti. It is not the dignity of craft which interests Pasolini: his hand-turned medieval items are stacked in cellars, not arrayed on studio walls, and are used to hide the loot or the illicit lover. *The Canterbury Tales*, more harshly photographed, reiterates the bawdy scenes of the earlier film with less lyricism. Sexuality, in Chaucer, is closer to sadism, witness the episodes of Alison and the smithy's poker, and Jenkins' nose. Women, less trammelled by Papal restraints, seem crueller and more wanton.

128

The most densely organized tale is The Miller's, in which two apprentices, bamboozled out of their rightful flour by the crafty miller, take their revenge by making love to his wife and daughter while he snores. Pasolini told Oswald Stack that neo-realism had never died, just moved from Italy to other parts of Europe; he cited Ken Loach's *Poor Cow* as an example of a late neo-realist film. But *The Canterbury Tales*, too, can almost be seen as a late example of neo-realism, not least in its use of street locations and non-professional actors.

In one significant way *Tales* dissipates the impact generated by *The Decameron*: the narrational device consists of Pasolini himself playing Geoffrey Chaucer, as he sits composing in his study, occasionally chuckling over a copy of *The Decameron* he has secreted in his desk, and being interrupted by his haggish wife at intervals. The vignettes occupy a less dynamic relation to the film's structure than the device of Giotto's assistant in *The Decameron*, suggesting that Pasolini identified less closely with the English text. In *The Arabian Nights*, the device of the narrator figure has disappeared, to be replaced by a more intricate pattern of scene organization. *Arabian Nights*, completed in 1974, was based on a 'critical reading' of the Scheherazade texts, according to Pasolini, though he admitted, in the film's press hand-out (published in *Il Tempo*), 'I am not all that well versed in Arabic history.' He added, however, 'There are some compensations. For instance, my existential knowledge of the Arab world. I can say that I know Arabs better

Nino Davoli in *The Decameron*

than the Milanese.' This assertion is characteristic; throughout his life Pasolini identified effortlessly with those who stood outside the Protestant work ethic and the Catholic matrix of sin, guilt and contrition.

What Pasolini grasped most thoroughly, however, in transferring his attention from Christian to Islamic culture, was the way in which self-perfectability replaced Christian ideas of social progress. This is perhaps what he meant when he referred to his 'existential' knowledge of the Arab world. Formally, the film reflects this, for it depends upon an interlocking 'box within box' pattern, characters and symbols interpenetrating the 'stories' until not so much a dramatic resolution but a point of stillness is reached, in the centre of the maze.

We might recall, at this point, that this attraction towards a self-referring rather than linear narrative structure is not entirely novel in Pasolini's repertoire. *Theorem*, for example, is organized as a game, and, writing in *London Magazine* (June 1969), James Price suggested that it was film-as-game: 'there are five players, and a joker,' proposing that each player has a chip – one basic character predisposition – and moves round the board until the joker interrupts the movement. 'On being intercepted they must leave the orderly grooves ... and try to cross the centre of the board, where there is a volcano.' The volcano is, of course, their own unconscious, a cauldron of neuroses, false values and inhibitions.

The setting for *Theorem* was a contemporary Milanese bourgeois household comprising industrialist father, ornamental housewife mother and two adolescent children, all of whom are transformed by the arrival of a saintly figure (Terence Stamp), whose 'chip' is perhaps his copy of Rimbaud's poems. Having seduced each member of the household, including the maid, he departs, leaving them to face their sudden change of values; each undergoes a flamboyant breakdown.

In *Arabian Nights*, the dissociation of the protagonists derives not from social alienation, as in *Theorem*, but from the workings of destiny. In this respect, the third film in his Trilogy of Life relates back to his earlier adaptions of Euripides. 'There is nothing natural in nature,' the centaur advises Jason in the prologue to *Medea*. 'Wherever you look gods are hidden. Myths are part of everyday life. The realistic man lives in myth.'

The Hellenic concept corresponds to the Islamic notion of destiny in *Arabian Nights*. In the sexual sphere, which is the film's main focus of human action, the haphazard coition of the first two films of the Trilogy is now mediated by erotic ritual, language and gesture. Infatuation, spiced with occasional gratification, replaces casual venery.

Pasolini himself has tried to elaborate the complex form of *Arabian Nights* – 'Every tale begins with an appearance of destiny, which manifests itself through an anomaly,' he has said, 'and one anomaly generates another ... so that mundane reality is confounded ... I wanted to approach the irrational as a revelation of life which becomes significant only if examined as a dream ... I have made, therefore, a realistic film, full of dust and the

The young lovers in *Arabian Nights*

faces of the poor. But I have made also a visionary film, where the characters are as in a trance, beset by an involuntary anxiety regarding the events
which surround them.'

The film opens with a caption which expresses this over-view more enig-
matically – 'The complete truth does not lie in one dream but in several
dreams' – hence the interlocking structure of scenes from different periods
of time and places. The first shot at once presents us with an anomaly: a
slave girl in an auction in the kasbah is *choosing* her new master. Only
later do we discover that the girl, Zumurud, has been knocked down to a
merchant named Rashid and her 'choice' of a handsome boy with 'innocent
eyes' is only a game. Following a night of dalliance with her virginal lover
she is reclaimed on behalf of Rashid by a menacing abductor, 'the man with
blue eyes', who seems to represent Christian imperialism, every entrance he
makes being accompanied by an ironic snatch of a European string quartet
which, in this context, becomes strangely threatening.

Throughout the film, anomalies of this sort recur; we see a marauding
caravan train pause to welcome into its midst a solitary vagabond instead
of cutting him down; we see two beggars bargaining over a job involving
the making of a temple mosaic – for *fewer* dinars not more; and we see a
monkey (recently a man who has had the misfortune to upset a demon)
writing elegant *pensées* – 'if you open the inkwell of power and grace make
sure your thoughts are filled with liberality.' The use of these anomalies
serves to sharpen the responses and reinvigorate the text.

In addition to its formal intricacy – the blend of melodrama, dream,
symbol and metamorphosis – *Arabian Nights* also presents us with a world
which differs intrinsically from medieval European communities, being matri-
archal *and* (therefore) phallocentric. The women in *Nights* collude to receive
the embraces of handsome boys, which are frequently received collectively

131

rather than individually. Women command, and women receive. What they receive, however, is not 'blood and gold' (like Chaucer's widow) but the 'supreme ecstasy' of semen, as Lou Salomé expressed it. When Uridin, Zumurud's lover, entrusts a merry matron with the delicate task of tracing his lost woman, she returns with a useful bulletin and takes for payment an intimate service. 'She who massages has greater value than the heads of wise men,' is a maxim cited by one jovial slave-owner.

The sexual passiveness of the males derives from the ineffability of certain ideas which permeate the world of Scheherazade. Destiny governs the lives of men, but so, too, does a fetching notion that 'All those with less beauty love the more beautiful,' an idea which is seen in action in the various affairs and infatuations which run through the film. The recurrent motif is the search for Zumurud by her innocent, somewhat inept lover. This is inter-sected by a variety of fables, including that of Aziz, who betrays his loyal fiancée Aziza with a sultry princess and, following Aziza's broken-hearted suicide, is castrated by a bunch of Aziza's vengeful women friends. What is interesting about Aziza's story is the curious complicity which develops between Aziza and the princess, who communicate in code language, using Aziz as a kind of erotic runner of messages between them. Aziza warns Aziz to observe the strict protocol of courtly love in his affair with the princess – 'She'll kill you if you make a mistake again.' To which Aziz replies, 'Fidelity is splendid but so is infidelity,' a remark he reiterates, to no avail, when the castrators seize him.

Other fables, too, hinge on betrayal and vengeance. A mosaic worker tells how he made love to a girl who had been incarcerated by a demon. The demon (played by Franco Citti with something of that eerie remoteness he also brought to the role of Ciappelletto in *The Canterbury Tales*) murders the girl and turns the man into a monkey. The love of another girl releases the beast from his simian condition, and he expresses his gratitude by completing the mosaic design in a temple. This design, containing a pattern of a deer and doves, echoes a prophetic dream of Zumurud's in which she visualizes her present separation from Uridin in the image of two birds trap-ped in a net, one of which escapes to freedom. Thus, dreams intersect, emblems are renewed, and culture continues, as the prefix to the movie suggested.

In the end, Zumurud, now masquerading as a 'king' in a strange desert city, and 'married' to a princess named Hayad (who knows Zumurud's secret), does become reconciled with Uridin; as they sink down together on the royal bed, an epigraph appears, 'The beginning was bitter but how sweet the end.' The sweetness of this resolution, however, depends upon a *formal symmetry* as well as an emotional repletion, and it is Pasolini's intuitive dex-terity in this, perhaps his most formally poised film, which is so commanding.

Andrew Forge, commenting on Islamic decorative patterns in *The Listener* (9th December 1971) reminds us that, 'A key feature of all Islamic art is that it suggests a precise meaning, even to eyes to whom that meaning is completely dark. A carpet of a faïence tile has about it something of the appearance of a text. It may not be clear where the script ends and decoration begins: the sense is that nothing is meaningless.' Forge suggests that a stern Islamic injunction lies behind the patterns of the artwork – 'God bids you beware of yourself ... unto him your journey is.'

Such ideas accord closely with Pasolini's design. In *Nights*, Aziz, castrated, is symbolized in the mosaic as the wounded dove, which echoes also Zumurud's dream of the trapped bird in the net. A similar effect exists at the end of *Satyricon*, when the principal players in the travelling fresco become arrested in Fellini's final frieze.

Forge also comments, 'The experience of looking at it [the Islamic artefact] comes as near as anything to the experience of reading, in the archaic sense of guessing or trying to make out.' This process of intuitive identification with the secret image is one with which Pasolini was preoccupied. In his essay on cinema-signs, which he named 'Kinemes', entitled 'The Scenario as a Structure Designed to become another Structure' (*Wide Angle* Vol 2 No 4), he considers the scenario as 'an autonomous technique, an integral work which may be *complete in itself*' [my italics]. He points out that the author of a scenario requires from his 'reader' a particular response, or collaboration, which consists of 'endowing the text with a visual completion which is absent, but to which it alludes. The reader ... quickly enters into a creative phase more mechanically elevated and intense than when he reads a novel.'

The 'hidden language' of the screenplay which Pasolini here postulates is evidently akin to the 'hidden order' of Islamic art, in Forge's diagnosis. *Arabian Nights* is Pasolini's most exhaustive realization of this hidden meaning, using the overlapping and interdepending effects of the fables. It is also his most paradoxical film, being almost entirely devoid of the conceptual schematization which mars, say, the Julian sections of *Pigsty* or the Toto/Davoli monologues of *Sparrows and Hawks*, while at the same time it does refer us to an inaccessible sensibility – that of the karma-dominated awareness of Islamic peoples.

Apart from its pertinence to *Arabian Nights*, Pasolini's essay on the scenario also reminds us that he was, throughout his life, a poet, essayist and novelist as well as *cinéaste*. In this respect, as noted, he bears affinities to Fellini, who has made excursions into prose and still retains his long-term interest in comic and cartoon drawings. An interesting illustration of Fellini's writing occurs in 1970, after he had completed the RAI TV documentary *I Clowns* (Clowns). Writing to his producer, Alberto Grimaldi, Fellini describes various projects he hopes to pursue, among them *Roma*, which, at this early stage, he calls a 'fantasy documentary' – Aldo Tassone later took up this term, abbreviating it to 'fantadocumentary', which he applied to all Fellini's hybrid movies in which autobiographical and documentary elements mix – *Notes of a Film Director, Clowns, Amarcord* and *Roma*.

Fellini wanted to convey in *Roma*, he suggested to Grimaldi, a young man's impression of the city, arriving in the late Forties poor and jobless. This perspective is partially retained in the completed film of 1972, but continues through to *Amarcord*, via the adolescent hero, Titta, who dreams of Rome while still living with his family in Rimini, during the Fascist era. 'Roman women have tits like water melons', Titta is told by older men, who know little more about the city than he does. Rome, in *Amarcord*, is a repository of provincial erotic fantasies. This has usually been Fellini's theme when considering Rome, his adopted city. As early as 1960, Roberto Rossellini noted that Fellini's view of Rome, as shown in *La Dolce Vita*, was essentially that of a provincial.

In *Roma*, however, Fellini takes care to show as well the modern reality – the throughways, the snarling traffic jams, the motorbike gangs which tyrannize the nocturnal city. In the sequence showing the car crash the camera equipment is also included, cranes, tracks and the massive rainhooded cameras, as though Fellini wants to suggest that Rome's film industry makes its own contribution to the general pandemonium. Babylon, to Fellini, implies not only the gratification of immense appetites but also indifference – dead souls in limousines making up their faces. In the middle of this Fellini himself, issuing orders through a megaphone telling his crew to follow a tourist bus – Fellini's ironic image of himself as the ultimate *paparezzo*, preying on the city's visual texture as those street-photographers prey on the tourists. In this respect, *Roma* anticipates *Orchestra Rehearsal*, which also attempts to show the relationship, however recessed, between pandemonium and the artist's search for authority and order. 'We all do what we're able to', Fellini tells a group of students in *Roma* who criticize his apparent lack of purpose. 'Now I want to reproduce life in a small variety theatre at the beginning of the war.' Offered documentary projects by Grimaldi, who may have been worried that Fellini's suggestions for future features after *Satyricon* were becoming too personal, Fellini proceeds to guy the format, using himself, through *Notes of a Film Director*, *Clowns* and *Roma*, as a nominal narrator figure.

The tacky variety show, circa 1939, which follows, in *Roma*, reveals a dancing girl, audience rowdies, a lame comedian, and an intellectual quoting Proust (and getting heckled) while someone urinates in the aisle and a dead cat is hurled on stage – 'cook it for supper'. A newsflash interrupts the action – 'The enemy attack on Sicily has been repulsed. Viva Italia.' An air-raid follows, and the theatre quickly empties.

While *Roma* has its surreal aspect, it also shows mundane views of Fascism. 'All this because of that old so and so', mutters one shivering drag artist in an air-raid shelter. Patriots among those huddled there berate the artist for 'defeatism'. A blonde lady shows her photos of husband and son, fighting in Russia. At dawn, as they all emerge from cover, someone remarks, 'They don't bomb Rome, because of the Pope'. Even in such dreary circumstances the notion of Papal Infallibility is refurbished.

Roma, however, adopts a flexible approach to the city's different phases of glory, taking us from the air-raids to the catacombs, where drilling proceeds on the (still) unfinished underground railway project. The scene showing the discovery of 'the shades' is as remarkable as any in Fellini's repertoire. The drills boring into rock are intercut with the painted faces of the frescoed walls beyond. The drills burst through into the chamber, and the colours of the frescoes fade in the sudden access of oxygen. The shots are notable not only for their cunning editing (Fellini's most striking visual effects are usually in composition rather than editing) but also because the scene was inauthentic – Fellini shot it at Cine Città. This stunning image of the conflict between modern predator and lurking relic has plenty of historical weight, however. In *Italian Cities* (Scribner, New York 1912), E. H. and E. W. Blash-field relate an authentic 'intervention' concerning the mausoleum of the Galla Placidia in Ravenna, where a conquered Roman princess who, 'for eleven hundred years ... sat upright in jewelled cerements in a sarcophagus', suffered a rude awakening in 1577 when 'some children, peering through an aperture in her sarcophagus, wishing to see better, thrust in a lighted brand, and she was burned, robes, cypress-wood chair and all, a strangely grotesque ending of this grim memorial.'

Parade of the Cardinals in *Roma*

Aldo Tassone, in his essay on Fellini in *Image et Son* (No 290), remarks, 'The spectator clearly understands it is a question of a dream which concerns him ... he sees the barrier that separates him from his buried ego fall and in an instant the enigmatic forms of the past ... come to light.' Fellini, however, has little time for the idea that, through his films, he achieves such clarity. 'The whole business about clarity seems to be some kind of aristocratic game,' he has remarked (*Fellini: Search for a Mythology* by Charles B. Ketcham, Paulist Press, New York 1976) '... All I want in my pictures is a real man who lives a real life, who worries about money, about his wife, about the Church and about his work.' This is a characteristic Fellini ruse, of course, designed to dissociate himself from the pretensions of critics, while suggesting a 'common man' image which is quite at odds with the 'aristocratic' films he makes. From the description offered above he might be describing a Comencini/Alberto Sordi comedy.

It is this contradiction, in both Fellini and Pasolini, between their 'street adolescent' self-image and their intellectual and artistic capacities which provides the rich creative tension. Pasolini also liked to dramatize himself as a rather more cunning version of his early hero, Accattone. His overt dissociation from the student militancy of the late Sixties is just one sign of this self-definition. Yet Pasolini could also admit, at the time of the Trilogy of Life, 'The only hope now is a cultural one, to be an intellectual ... for the rest I am consistently pessimistic.'

However strenuously Fellini likes to deny his own talent for analysis, his essays on provocative cultural issues such as censorship or the impact of television reveal his conciseness. As one example, consider his remarks on the implications of censorship: 'It is a way of admitting our own weakness and intellectual insufficiency.' His talent for self-analysis is also evident, in 'Maybe I have survived because I have a fakir's wall around me. For others it is not so easy. They don't all have my fanaticism.'

Fellini, in *Amarcord*, adopted the form of the annual cycle, like *1900*, *The Tree of the Wooden Clogs* and *Christ Stopped at Eboli*. The importance of this structure to *Amarcord* is a topic on which Fellini, in his 1967 essay on Rimini, speaks emphatically: 'There is a sharp division between the seasons at Rimini. It is a substantial change, not just a meteorological one, as in other cities. There are two quite separate Riminis.' In the film, however, the intermediate seasons, spring and autumn, are just as distinctly visualized. The film opens, in fact, with the ceremony of burning the winter goddess in the town square, as the puffballs drift in: it ends on the same image of puffballs, after another important ceremony, a wedding.

The inspiration for *Amarcord* derives from an earlier project Fellini particularly wanted to realize, which would have been called *The Unknown Woman*. In the letter to Grimaldi in February 1970 (transcribed in *Positif* No 200), Fellini alludes to three movie projects – *Roma*, and two ideas for feature films, one of which is *The Unknown Woman*, the other being *City of Women*. The former he describes as 'a metamorphosis ... about a man who can no longer stay with a woman because he has projected on her every-thing. When this woman quits him ... he works backwards, looking for her in all the places she has lived from birth, locks himself in the villa where they lived, becomes a fetishist, begins to change into her ... the Frankenstein monster in the grip of a delirious love ... he has become her.'

This plot outline, which somewhat resembles Polanski's *The Tenant*, sounds unlikely as a Fellini vehicle. His films have always been baroque in *form* rather than content. Grimaldi must have thought so too, for Fellini next made another 'fantadocumentary', not *The Unknown Woman*. What is especially interesting, however, is the description Fellini offers about the style of *The Unknown Woman*. He had it in mind to shoot in a Northern town, probably Mantua, 'amid the fogs on the motorway ... the lights, the silence, the colours, the unseizable images. Shadows of cars, gleaming phan-toms.' The film would be set in 1928, and would deal with Italy's deep cultural isolation during those years of Mussolini. On this theme Fellini became unusually eloquent: 'Fascism – imbecilia, erotomania, cowardice and masochism, the clergy, lugubrious carnival masks ... the funereal erotic-ism of D'Annunzio, the warlike temerity of his enterprises, which reveal his paranoid nature.' He further enticed Grimaldi with suggestions for a snatch of the *Mille Miglia*, a crushed dog on the road, a lyric opera sequence, and above all the importance of the fog, 'which deforms everything'.

In *Amarcord*, many of these ideas have been incorporated into the film's texture – the fog, the motor race, the Fascist carnival, the erotic obsessions of the menfolk.

Amarcord, like *Satyricon*, provoked critical disapproval. Edward Murray, for one, commented, 'Fellini has merely provided a mechanical external scaffolding. There is no doubt, no tension, no coherence ... *Amarcord* is barren of much that is good, true and beautiful in human experience.' Such critics of Fellini's later work are the 'Gelsominaphiles' who reprove Fellini for not continuing to make films with the crisp moral polarization of the earlier 'mystical waif' movies, *La Strada* and *Nights of Cabiria*. Yet such an idealized vision of human relations could hardly be maintained through the years of maturity. What Fellini strives for in the later feature and TV films is a cinema of poetic reinvocation.

Amarcord, based on his essay of 1967, reconstitutes the vanished world of Fellini's childhood, while it reveals the essential mindlessness of Fascism, as it existed in that town during the early 1930s. The earthiness of *Amarcord*, which has so scandalized some critics, can be seen as Fellini's attempt to disengage *his* world, the world of Titta's turbulent extended family, from the spiritual sterility implied by Fascist domination. Fellini, however, unlike Bertolucci and Pasolini, does not make his case against the Fascist era in strident terms. Instead of the atrocities repeatedly shown in *1900* and *Salò*, Fellini shows just one incident which illustrates the imbecilic violence of small-town Fascist officials. Titta's father, Augusto, who defies the authorities on the eve of a Fascist rally by planting in the clock tower a phonograph which grinds out the *Internationale*, is arrested and 'interrogated'. The torture inflicted is the relatively mild, but humiliating, castor oil treatment. Augusto is then released, while an official complains, 'Why can't they understand that we Fascists only want to give them a sense of dignity.'

In the scene following Augusto's interrogation, Fellini shows him at home, in the family hipbath, being sponged clean of his own ordure by his wife. For once the repetitive quarrel between them is dormant. Titta, sneaking in for his usual giggle at father's expense, is delighted, and quite oblivious of the scene's implications.

In this way, with none of the melodramatic indignation of Bertolucci, and none of Pasolini's insistence on the 'blessed rage for ordure' (to paraphrase Wallace Stevens), Fellini quickly shows what daily life was like, growing up as a non-sympathizer under Mussolini. It is a measure of the poetic scope of *Amarcord* that Fellini later shows the reality of Fascism by use of a visual metaphor, as grandfather and Titta grope through a dense fog that envelops the town, calling 'Where am I? I'm not anywhere. If death is like this ... no birds, no wine ... up yours!' Fascism, which blots out joy and spontaneity and eliminates trust and loyalty among men, is now expressed as visual disorientation, that element which the cinema, of all the arts, and Fellini, of all film artists, can so effectively convey.

Earthiness, in this context, can be seen not as a coarsening of Fellini's vision, as Edward Murray would have it, but as a vital defence against the repressive hygiene of life in an authoritarian community. It is no accident that Titta's family is always fighting. How else can they relieve the tension? And where else but over the dining table? Thus *Amarcord*'s ribald stressing of bodily functions can be seen as a form of healthy defiance.

Amarcord also touches lightly, but subtly, on other eternal human issues. Madness was one form of escape for those unable to endure the moral, emotional and cultural deadness of the Mussolini era. Titta's uncle Teo, whom we meet in the memorable picnic sequence, is a shining example of madness as a shrewd defensive posture. 'He was the most intelligent of us,' says Augusto sadly, collecting his brother Teo for his Sunday outing to the country. Titta knows no more about his uncle than his family has told him. Standing beside Teo on the edge of a cornfield, he sees a man much taller, and calmer than his father. His first intimation that Teo is different occurs when he realizes that Teo has not bothered to unzip his pants as he urinates, so taken is he with the 'blue streak' on the horizon.

At the farm where the family lunches *alfresco*, further signs of Teo's oddness manifest, when he climbs a tree, shouts 'I want a woman' and resists all attempts to persuade him down. The truth of Augusto's remark that Teo was the most intelligent of the family emerges when they try to outwit him, creeping away as though leaving. When Augusto pokes his head round the farm wall a stone lands on it. Teo is still one jump ahead, as madmen usually are, Fellini suggests.

Teo's desire for a woman is 'heard' in typically ironic Fellinian fashion – a diminutive nun comes to order him down and march him back to the institution. But what Teo dares to express as an open demand, Titta, also sexually hungry, has to deal with furtively. In a small town, there are precious few outlets for lads of Titta's age who seek the basic connection. Fellini's most explicit visualization of this group-frustration is the vignette which shows the gang of boys crammed in an old car which is stationary, but wildly rocking. The men are no less frustrated in Rimini, however. Even Vulpina, Fellini's latest in a long line of hypersexed females (one recalls Saraghina in *8½* and Oenothea in *Satyricon*), cannot satisfy the inflamed appetites of the male population, as she slinks by the building sites, stroking herself in the full gaze of the workmen. Vulpina's sexuality is so ferocious that she scares the men, who tease but never touch her.

Perhaps because of his female vamps, Fellini has been characterized, in his own country at least, as a misogynist. In the film magazine *Cinema Sessanta* (No 95, February 1974), in an article 'Fellini e la donna', he rebuts the charge – 'But I don't hate women. I love them very much,' he says. Fellini has elaborated his views on the subject more fully elsewhere: 'Women are stronger because they are nearer to nature,' he has asserted. 'Intellectuals are lost in pools of anguish for the future. Not the woman – she is here now.' Edward Murray has aptly observed that the role of women in Fellini's films is vital to their inner coherence for they 'frequently challenge the man to overcome his narcissism'. This nature versus nurture axis to Fellini's perspective on men/women relations is marked, of course, in *Juliet of the Spirits*, which offers the most comprehensive view of the sustaining power of female modesty and intuition that Fellini has yet undertaken.

Fellini, of all the Italian film-makers today, has retained a memory of his Roman Catholic origins (what Antonioni has called his 'authentic Catholic nostalgia') which enables him to examine sexual relations in terms of coyness and prurience as well as expressive behaviour. Titta's sexual turmoil in *Amarcord* arises from the tension between his upbringing – 'the saints cry when you touch yourself,' the priest scolds him – and his yearnings. Pasolini also praised the physicality of *Amarcord* when he remarked, 'the poetry is bound up with the body of the characters' (*Playboy*, February 1974). Fellini's apparent misogyny is, therefore, no more than an aspect of his distance from commonplace material obsessions, which leads him to a fascination with those forms of behaviour which follow 'a different drummer'.

Amarcord ends on Gradisca's wedding, which takes place in the meadows outside town, as once again the puffballs drift in the spring breezes. She has married, not her handsome Ronald Colman, but a portly naval officer. 'Remember your duties, and have lots of babies,' is the town's final counsel to her. In his essay on Rimini Fellini describes how he went looking for Gradisca, thirty years later. He drove to 'a wretched little village' where he saw 'a little old woman hanging out washing in a garden ... "Can you tell me where Gradisca is?" "I am Gradisca," said the old woman ... she had lost every trace of that carnival glitter of hers.'

Fellini has always been responsive to such temporal messages. In *Amarcord*, he rebuilds the Rimini of his childhood, distilling its essential characteristics, bequeathing us a portrait of a town, and an era, unparalleled since Dylan Thomas's *Under Milk Wood*. Fellini's phenomenal Rimini stands defiantly beside the real town of today, which Fellini dismisses as 'fifteen kilometres of nightclubs and illuminated signs ... a kind of milky way made of headlamps ... the night has vanished.' He adds, regarding his own feelings when visiting new Rimini, 'I felt foreign, defrauded, diminished. I was at a party that was no longer for me. At least, I no longer had the strength and greed to take part in it.'

Rimini, in *Amarcord*, is the 'eternal province' of childhood, in which every detail stands out clearly, even in the fog, precisely because it has been experienced by the as yet unsplintered concentration of the child. Aldo Tassone praises Fellini for not depending in *Amarcord* on his own presence before the camera, as he did in *Clowns* and *Roma*. But Fellini has no need to represent himself in *Amarcord*; he is still at the centre. It remains to ask one question: does *Amarcord*'s evident attachment to the past bear out Fellini's own remarks about his sense of fatigue during the Seventies? Would a man immerse himself so deeply in the Thirties if he knew how to confront the present era? This question is best considered in the context of Fellini's two 'blackest' projects, *Casanova* and *The Orchestra Rehearsal*.

The very absence of any social ideal which had so struck Fellini when considering the implications of the hippie movement in the late Sixties now suffuses his own attitudes, as they are synthesized in *Casanova*, his most vexing production to date, both artistically and financially. 'Casanova is a completely external man,' Fellini told Aldo Tassone (*Positif* No 181, May 1976), 'without secrets and without shame ... a presumptuous man, a know-it-all ... he has gone all over the world and it is as if he has never got out of bed. He is really an Italian: the indefiniteness, the indifference, the commonplaces, the conventional ways, the façade, the figure, the attitude.'

Casanova occupies a special place in Fellini's repertoire, being the only one of his films in which he is entirely out of sympathy with his subject. In this respect, it raises questions concerning Fellini's relationship with his leading actor, Donald Sutherland. Fellini's choice of the American actor was based on his notion that no Italian actor had the physical presence required of the part: Casanova and Sutherland, at around 6′ 4″, share an affinity of height, if nothing else. Fellini's desire to create, with his leading actor, a repellent physical type must have involved a director/actor complicity as subtly manipulative as any Fellini had undertaken. Commenting on his relationship with the director, Donald Sutherland has observed, 'I was caressed and cuddled and loved and treated more gently ... than ever before

Donald Sutherland in *Casanova*

in my life' (*American Film*, September 1976). The result of this oddest of all Fellini's curious 'liaisons' with his leading artists seems to vindicate the choices made by actor and director. Sutherland's performance, which blends the high-minded opportunism of Casanova with his narcissistic erotomania, is that very rare event in screen acting – a 'star' performance which is entirely underivative, a truly definitive performance.

The making of *Casanova*, besides the curious liaison with Donald Sutherland, contains other aspects of Fellinian dada. During script preparation, for example, Fellini would tear each page from Zapponi's early edition of the *Memoirs*, to the dismay of his co-writer. He also took a break while shooting to visit a medium in order to 'communicate' with Casanova, who proved, according to Fellini, just as bumptious in the next world as he had been during his earthly round. Casanova's shade had the nerve to offer Fellini a morsel of advice on the taking of carnal pleasure – never before dinner or standing (*American Film*, September 1976). Thus Fellini, via a series of ruses, 'discovered' a movie after all in the self-congratulatory musings of the *Memoirs*.

'Eighteenth century Venice,' wrote James Morris in his book (Faber 1960), 'was a paradigm of degradation. Her population had declined from 170,000 in her great days to 96,000 in 1797, though the Venetian Association of Hairdressers still had 852 members. Her trade had vanished, her aristocracy was hopelessly effete, and she depended for her tenuous existence upon the good faith of her neighbours.' Giacomo Casanova, a man in middle years when we first meet him in Fellini's movie, is very much a man with his values formed by these conditions. This 'turd of a man' (as Fellini calls him) is the eternal opportunist, who preys on the fear and vanities of those around him and who always keeps moving, a Wagner figure without the stabilizing influence of a Ludwig the Second in his background. Casanova's activities across Europe, as tracked by Fellini/Zapponi in their screenplay (with additional material supplied by Tonino Guerra and Anthony Burgess), begin in Venice, where we see his arrest and subsequent escape from prison, and also end there, but in between range across Northern Europe. Although Fellini is obliged to be interested in the sexuality of his hero, he is actually more conscious of his charlatanism, an important aspect of the Venetian inheritance. Indeed, one of Casanova's earliest memories was of the witch's incantation chanted over him during a childhood nosebleed.

The charlatanism of Casanova is shown to be his dominant characteristic; the hyperactive sexuality is an element within this broader area of self-seeking. A brief account of the film's narrative reveals the way in which these different forms of opportunism mingle. In the first sexual encounter of the film Casanova's presence is requested at an island convent. Here Casanova performs vigorous coition with a provocative nun (or whore in nun's clothing). The coupling is watched by a voyeur hidden behind a wall on which is painted a fish – the eye of the fish is his peephole. But when Casanova announces that he expects letters of introduction for his services just rendered the 'eye' is vacant. On the way back to Venice, Casanova is arrested by officers of the Inquisition, and summarily imprisoned.

Having escaped from prison over the rooftops of Venice, Casanova arrives at the salon of the Marquise D'Orfé, where he tussles verbally with a waspish count and takes part in a séance. The Marquise has a special mission for him and takes him to her inner sanctum, which contains the alchemical works of Paracelsus. Casanova is to perform 'a great work' for her – once again, this turns out to be a euphemism for copulation.

Casanova enlists the ancillary services of a nubile girl named Henriette, who provides vicarious stimulation 'from the wings' while Casanova forces himself through the *'conjunctionem'* with the elderly marquise. From here he moves on across Europe, the impetus arising from the conquests of different women. As Gilbert Adair remarked, reviewing the film in *Sight and Sound* (Autumn 1977), 'Single-mindedness becomes one of the formal components of the film.' Casanova's travels seem to be the message.

A tension is established, however, just as a tension existed in *Satyricon* between the 'single-mindedness' of the protagonists and the 'multiplicity' of Fellini's vision. While Casanova is intent only on the main chance, whether it be erotic or financial, Fellini binds us to a series of decors and pageants which serve, through the carefully differentiated textures, to counterpoint the hero's self-importance with a world of proliferating egocentricities. Thus, while Casanova himself is always unctuous – 'women are sweeter and more reasonable than men,' he brays, 'we are born to oppress them' – the latent barbarism of these sumptuous European courts is allowed to emerge. Casanova's couplings, and tired aphorizing – 'what is a kiss? Simply the desire to immerse yourself in the soul of the woman you love' – is shown to be part of a *generally* callous rhetoric.

What Fellini objects to particularly is the devitalization signalled by these cancerous courtly appetites; their sheer excess is evidence of a spiritual blight. This is, of course, his essential theme, equally plain in his studies of ancient and modern Rome. Fellini is an advocate of modesty and common sense; he lives simply; his own values seem similar to those of Rossellini, with whom he worked in the post-war years as scriptwriter and assistant director. Given the clash between the worlds Fellini likes to depict and his personal domain, a satiric impetus is inevitable. *Casanova* is a film about narcissism, morbidity, exploitation and bombast. Casanova is a man defined by his own pretensions. He sees nothing of the world around him. He is commonplace, his senses are closed. Where there is no humility, Fellini knows, there can be no development of self. This intact core of the Rossellinian spirit of neorealism Fellini retains today. It is the source of his artistic vitality, even while he claims to suffer a diminution of energy.

It is this inner equilibrium which lies at the heart of Fellini's creativity, suggesting a parallel with the British writer J. C. Powys, whose artistic predispositions had so many affinities with Fellini's. Powys, too, lived simply, possessed great gifts of visual imagination and a vivid sense of the numinous, expressed in his novels via his 'eidolons' (guiding spirits of nature). Both share, too, a sensitivity towards the world's holy fools; compare Fellini's attraction to itinerant waifs, Gelsomina and onwards, to Powys's similar invocation of the circus girl in *Maiden Castle*. Above all, both artists protect their exceptional sensitivity by a variety of ruses. Powys embraced the English defence, self-laceration – which he called his 'malice'. Fellini, preferring what he calls *cinéma-mensonge* to *cinéma-verité*, has confused and provoked those who seek to analyse him.

Such plasticity of mind is sustained by coherent instinct rather than ideology, whether political or religious. Hence the seeming apoliticality of Fellini, his freedom from cant, his insistence on 'doing what he wants' as Henry Miller, praising *Satyricon*, put it. Fellini's instinct tells him that Casanova is a hollow man. 'Your travels take you through the bodies of women, and that leads you nowhere,' Casanova is told by one courtier, who might serve as a mouthpiece for Fellini.

Each coupling, in fact, is designed by Fellini to exert the maximum satirical effect. At Berne, in the house of a Doctor Moebius, Giacomo meets his match in the Doctor's alert daughter, Isabelle, who gulls him into a rendezvous at a hotel, where she has engineered a small orgy for him, involving a movable bed and an insatiable hunchback girl. The following night, in Dresden, Casanova attends the opera and accidentally meets his aged mother.

This encounter is beautifully mordant, for Casanova's egocentricity is for once dwarfed – by his mother's. His unctuousness, deprived in this context of any pragmatic motivation (for his mother is the one person he cannot fool), becomes a mannerism; he seems querulous, his charm falls on deaf ears. When he takes his leave, outside the grace and favour villa she has comandeered, he seems more pathetic.

The most revealing of all the sexual episodes occurs at the Court of the Duke of Würtemberg (played by British actor, Dudley Sutton). Here Casanova offers his services as ambassador, while the mad Duke sets in motion a synaesthetic happening of which Scriabin himself would have been proud – six mighty pipe organs clash simultaneously in the great hall. Casanova, drawing his sword to protect a maiden from assault by court churls, discovers she is a mannikin, clockwork driven, but uncannily womanly in tint and curvature. At this point Fellini's cold-eyed satirical gaze achieves a pinnacle. Casanova is fascinated by the mannikin. That night he creeps down and 'makes love' to it. Never before has he given himself so fully or spoken so tenderly towards a lover. In this onanistic act, he achieves perfect oblivion; he has found his ultimate erotic frisson.

Having trounced his 'enemy' in such a definitive scene, there is little left for Fellini to do but despatch him. This he does, with a large elision in chronology, jumping forward in time to the Court of Dux, in Bohemia, where Casanova resided as court archivist in his declining years. The pompous philanderer has now become a petulant windbag. When he goes to complain about his poor treatment to the elderly aristocrat who reigns supreme, he is met with the same massive indifference to his problems that he previously encountered in his mother. 'There's a crane on the roof of the woodshed,' she points out, ignoring his litany of grievances. Later, his portrait is torn from a prized book and pinned to the lavatory wall with ordure. The servants serve him anything but his favourite macaroni. When he reads his poems, the young folk giggle. So the old ruffian, in his peacock ruffles, with his boiled expression, retreats to his boudoir where his few mementos of past sensuality are gathered – the music box he always wound before seductions, the dressing-gown, the pen and papers. 'Will I ever see Venice again?' he frets. Back we go, to the ice on the Lido, watching again that spectacular opening sequence as the queen of ice rises from the lagoon, lit by flares and fireworks (this image, Fellini has said, has haunted him since he first dreamt it as a child). Faces and silhouettes of Casanova's past conquests flit briefly through the scene. A winter wind moans, and finally the camera closes on the manikin from the Würtemberg palace, revolving slowly on the ice, a sexual mechanism, Fellini's own verdict on his hero.

'Casanova has been described as a psychological type of instability,' noted Havelock Ellis in 1911. 'That is to view him superficially. A man who adapts so readily and effectively to any change in his environment or in his desires only exhibits the instability which marks the most intensely vital organisms.' In the context of Fellini's crushing verdict on Casanova, Ellis's comment seems highly pertinent. Has Fellini finally been confronted with a form of vitality so intense that he cannot recognize it for what it is? Has his provinciality, confronted with his 'enemy's cosmopolitanism', finally become an impediment? Whatever we decide, and perhaps only Fellini's films during the Eighties will help define these questions for us, one conclusion can be reached. *Casanova*, while it may be highly subjective as a view of this Enlightenment figure, does present very clearly Fellini's way of handling a subject with which he is at emotional odds. He was able to participate fully in the project only by introducing a tension between the historical figure and his own disgust for him.

One theme from *Casanova* links directly with *Prova D'Orchestra* (*Orchestra Rehearsal*) of 1979. The acoustic cacophony of the Würtemberg organs is a motif taken up and made central in Fellini's RAI TV movie, again strongly satirical in its depiction of a riotous musicians' strike during an orchestral run-through which is due to be televised. The satire comes in two phases in this seventy-minute study. First, the trivial bawdiness is exposed as the musicians prepare for the rehearsal. A trumpeter bursts a contraceptive on the bell of his horn, to much laughter. Several women musicians stress the sexual connotations of their chosen instrument – the violinist, for example, finds her fiddle 'molto phallaccio'. The winsome young pianist flirts with all and sundry. Someone gets out his stopwatch, someone else gets out his pasta. As the harpist arrives the percussion escorts her to her place with a sly *tarratiddle* on the cymbal. The older males read their sex magazines, or turn to the football match on their transistor radios.

Then begins the television interview, in which Fellini is heard (but not seen) asking the musicians about their instruments. What emerges is not so much individuality of choice as narrow egocentricity. Each reckons his own instrument is the most important and justifies his choice in subjective terms. The initial interviews ending, further fractiousness develops. Cellists fight over a music stand, a rat bolts across the floor of the crypt and is hounded to extinction below the etchings of Mozart and Schubert. The autocratic conductor arrives, and is angry when the music (a suite by Nino Rota, Fellini's long-standing musical collaborator) sounds ragged. After several more attempts the conductor calls a break. During this interval, in which he talks backstage to Fellini's camera, the satire is intensified. Triviality and bawdiness now cede to bloody pandemonium in the ancient crypt – fornication and rioting, with boogie from the keyboards. Two young brass players scrawl graffiti on the walls while the conductor (Balduin Baas) reminisces in his dressing-room over warm champagne – ice cubes, like harmony, are part of a vanished order, he informs us, recalling happier times in Tokyo and America, while dabbing on eau de cologne.

Fellini stiffens his vision of chaos with a Ferrerian touch of the apocalyptic, in the form of a strange boom which recurs, ever louder, shaking the ceilings and fusing the power supply. As the conductor primps by his mirror and prepares to return to the fray an orgy commences. The music is jungly.

One militant bellows, 'Music is an instrument in the class struggle' and hauls the pretty pianist under her piano, where she munches her hamburger as he lies on top of her. Fights break out. A vast metronome is marched to the dais as one elderly string player produces a revolver from his sock and fires wildly around him.

Then the wall caves in. A demolition ball and chain hangs in the hole, like a view of planet earth from space. One casualty looks fatal, and during the silence of sudden dread the conductor reasserts himself. 'Why are we here? Back to the instruments, please. We are artists. We must continue.' The film ends on this resumption of the runthrough, in the ruins, while the conductor rants in German.

Fellini has said that he first conceived *Orchestra Rehearsal* while attending recording sessions of his film scores with Nino Rota. 'I was always surprised to see the emergence of a unique form,' he has commented (*Postif* No 217). 'This process of making order from chaos provoked a strong response in me.' His film, although it seems to be doing just the opposite, making chaos from order, is not merely a perverse caprice, but is based on research Fellini conducted with many musicians, from different regions of Italy. 'I managed to take from each a little of that madness with which each identifies with his instrument,' he has said.

The professional inter-rivalries which Fellini exposes in his film satire have in fact been researched in this country, too, by John Davies, in an article he wrote called 'Orchestral Discord' (*New Society*, 8th January 1976). Davies notes that musicians in a famous British symphony orchestra have 'strongly stereotyped ideas about each other'. He reports that the brass playing stereotypes (as seen by string musicians) comprise 'slightly coarse and unrefined heavy drinkers' while the strings are seen by the brass as 'like sheep, precious, touchy, physically rather delicate'. Davies concludes, appositely, in the light of Fellini's satirical movie – 'Now, during a Mozart performance, seeing the trumpet players sitting devout and serious at the back, one wonders whether they are really reading *Playboy*.'

Fellini's theme has also been argued by Hans Keller, who declares that 'the orchestra has started to disintegrate, and that in proportion ... players have assumed personal responsibility for their phrasings.' Like Fellini, Keller sees the collapse of the principle of collective harmony. It seems, then, that Fellini's movie may not only have the metaphorical significance that certain critics have attached to it, but also quite literal substance. Fellini himself, however, was interested in the ethnic clashes occurring in his movie as well as the musical. 'I wanted to discover the pathological sense of each region,' he informed Michel Ciment, though in practice, due to budget factors, most of the musicians appearing in the film come from the Neapolitan region of Italy.

Although Fellini was willing to amplify the theme of destructive competitiveness from the musical to the ethnic level, he carefully avoided any political interpretation. 'For some, in Italy, it's a film about the Historic

Players and Conductor in *Orchestra Rehearsal*

Compromise, for others a mystical film, for others a conservative film,' he stated. He prefers to regard *Orchestra Rehearsal* as a moral tale about the individual's responsibility to attain his own form of coherence. 'If one abrogates the responsibility for one's own life,' he has commented, 'there's the danger of falling into collective indifference. For me, it is necessary to be the father of oneself.' Political analysis can only devitalize, he argues: 'Politics is the reduction, the enfeebling of life ... a political judgement on a film prevents an emotional and individual impact.' This is a crucial element to Fellini's thoughts about cinema. Certainly some British reviewers have criticized the film's absence of politics. Derek Malcolm, for instance, referred to Fellini's political 'naïvete' while Philip French decided that the film was 'too simplistic and smugly assertive'. Oddly, certain Italian politicians have reacted more intuitively towards the film than most of the reviewers. Former President Bertini, for instance, noted after a preview, 'This film is neither progressive nor reactionary. It's true.'

What is most urgent about *Orchestra Rehearsal*? Is it simply a film about urban stress, and the retreat from individuality to solipsism? Is there, on the other hand, a more precise intention? Perhaps Fellini wishes to show how critical pressure is exerted on a small, contained community – the orchestra – by a TV inquiry. Lorenzo Codrelli, in *Positif* (No 217), has suggested that the TV inquiry turns the musicians into exhibitionists, provoking their greed for performance fees, and their subsequent frustration. Visually, Fellini confirms this view; it is no accident that the man who really looks as though he is in charge of the proceedings is not the conductor but the imposing union man, who sits by the telephone. In addition, the conductor's own self-involved volubility during his backstage interview could be seen as that

147

very absence of interest in the orchestra which helps to induce a riot.

Implied, then, is a wary antagonism towards the investigative role of television, a theme which Fellini has addressed directly in his essay on television in 1967 – 'the sacred aspect of the spectacle is missing' from TV viewing, he noted. Cinema, being communal, serves to stabilize an audience, while TV, fragmenting the viewer with a plethora of images, only further undermines his equilibrium.

It is significant that, in *Orchestra Rehearsal*, the only musician who stands apart from the mob frenzy is the harpist, and Fellini has described the harp as 'the only instrument capable of giving a feel of other dimensions' (*Il Tempo*, 24th February 1979). It is the 'other dimension' which television fails to glimpse. He also remarked, following the press show in Rome in February 1979, 'Instead of showing politics on TV, let us show the information of our unconscious. The film speaks of the danger of moving away from the unconscious to the collective superego, which is politics.' It is not often that one hears an Italian film-maker dare to suggest that Italian society is perhaps over-politicized, as Fellini here hints.

Since *Orchestra Rehearsal* was to prove Fellini's final collaboration with Nino Rota, who had composed the scores to every one of his movies, it seems fitting to make brief reference to Fellini's musical relationship with Rota. Although Fellini has said that music 'creates in me a sort of mistrust' (*Positif* No 217), he wrote, in his obituary for Rota (*Il Messaggero*, 13th April 1979), that Rota was 'like a child ... he said brilliant things on the spur of the moment ... it was as though he did not see my films. He had a geometrical imagination ... he lived inside music.' Since Rota's death, Fellini has completed *City of Women* and has talked of re-using existing Rota scores, not only in this film but possibly in others which may follow. At this juncture in his career, there seems little slackening in his artistic vitality, for all the nay-saying which his films of the Seventies have elicited in certain quarters.

From Fellini's account of Fascist Italy in the early Thirties to Pasolini's depiction of the Salò atrocities is a broad leap, but a leap with some symmetry, for what Fellini, in *Amarcord*, shows beginning, Pasolini shows as it ends, when Mussolini's regime collapses. The facts of Salò are grim. The republic of Salò was the last bastion of Fascism during the final eighteen months of the war. Here, during the reign of terror that was instituted, 72,000 were murdered, 40,000 mutilated and roughly the same number sent to camps. During the most notorious episode of the regime, 2,000 villagers were massacred at Marzabotto and a number of women and children were sexually humiliated. Salò is not an episode in recent Italian history the people of Italy wish to remember. Yet, when Pasolini approached the subject, he introduced a stroke of icy inspiration, cross-fertilizing the facts and figures of the recent historical atrocities with an adaptation *en bref* of de Sade's massive pornographic text, *The One Hundred and One Nights of Sodom*. The work was written at speed while the Marquis was in prison; it is his catalogue of perversions, from bestiality to necrophilia. Pasolini, in adapting *Sodom*, elided the worst of de Sade's sexual atrocities, whereas when he adapted Boccaccio's text he chose the most salacious episodes.

Revitalized by his success with the Trilogy of Life, Pasolini felt able to approach the challenge of *Salò*. His decision to shift the chronology of de Sade's work from the late eighteenth to the mid-twentieth century was an elegant one, given the fustian quality of the work.

The relocation of the dream locale of Sodom to the puppet republic of Salò, in Northern Italy, enabled Pasolini to give the text a political character. Pasolini's older brother was killed during the resistance to Mussolini's republic in 1944. His choice of historical context was quite unambiguous.

Final moments of tenderness in *Salò*

Perhaps what has most disturbed the European critics and cinema public about *Salò* has been just this *absence* of ambiguity about Pasolini's final work, his most formally perfect movie. *Salò*, while it is one of several recent Italian films depicting the atrocities of Fascist regimes, is, more than any of these, a *baseline* study of Fascism, showing a sexual, cultural and social malaise as well as a purely political degeneration. The four 'masters' of the Hall of Orgies, Priest, Advocate, Politician and Aristocrat (based on de Sade's own authoritarian figures, Duc, Monsignor, Ambassador and President), are not individualized but represent types of arrested development. In a sense, they are the bogeymen of classic psychoanalytical abstraction, showing various forms of infantilism, with their anal and coprophiliac fixations. *Salò* is, therefore, a more densely abstract profile of the Fascist mentality than, say, Bertolucci's portraits of Manganiello in *The Conformist* or Attilio in *1900*.

Salò begins with a wide angled pan across the peaceful lake at Salò, as seen today. The vista is peaceful: a few soldiers hanging around, and then the signboard. An opening reminiscent of the equally unemphatic first shot in *Oedipus Rex* (following the prologue), a signpost pointing to Thebes, that other small oligarchy where man plumbed the bottom. Inside a conference room at a requisitioned mansion four middle-aged men are signing a 'constitution'. The reign of terror is formally ratified. It is 1944. Road blocks. Round ups. Mothers crying. Children hiding. This is the 'prologue'. It is called 'anti-Inferno'.

The young victims of the round-up are taken to the commandeered mansion which Pasolini sees as the inferno. He divides the remainder of his narrative into Dantean 'circles'. The first is the 'circle of manias'. Now we meet those crucial 'aides' to the Masters, the four former brothel madams, whose role is to eroticize all future proceedings: one of them plays cocktail tunes on a piano, the others in turn tell pornographic stories and attempt to wheedle or intimidate their victims into a semblance of sexual arousal. Initially, the madams require some training by the masters. One is told, after starting a story, 'You must not omit details. I didn't hear anything about the shape of the penis, or whether he touched her.' The blonde madam continues while a youth is dragged into a sideroom by one agitated master, who emerges alone complaining that the youth was no good. The madam offers her services, which are declined. 'I'll wait,' says the master. In this world of twisted sexuality, gloating is often preferable to the act itself. There follows a lesson in penile arousal from a madam instructing a bunch of frightened girls. 'The men are not satisfied. The first thing to learn is how to hold it.' A girl is made to undress a manikin and masturbate it. Like Fellini, Pasolini has understood that the sexual facsimile may be more arousing to the sexually obsessed than a living body. The scene ends on a shock cut to a girl whose throat has been slit.

The 'circle of manias' concludes with a grotesque mock-wedding, in which the 'bride' and 'groom' are forced to lie down together amid a pile of white lilies. They are ordered to strip and perform. The scene, like many in the film, pulls two ways at once; beneath the terror of the young couple there is a moment of authentic tenderness as they start to make love. A master rushes forward to stop them – 'that flower is reserved for us.' To the masters, the moments of real emotion shown by victims are the obscenity.

While a madam proceeds with a tale the masters discuss the aesthetics of violence and force the victims to eat *policella* on the floor, like dogs, all this accompanied by the ubiquitous cocktail piano. One girl sobs that she cannot go on. Two youths slump together; one masturbates while the other doodles listless patterns on the floor.

'The circle of shit' begins as a madam sits at her dressing table checking her make-up. A bomber drones overhead. She descends the stairs to the Hall of Orgies and tells a story about killing her mother. 'I sped her into the next world. I never felt such a subtle pleasure.' There are echoes here of Clementi's 'I have killed my father and felt exquisite ecstasy', which he re-iterates like a mantra as he faces his executioners at the end of *Pigsty*.

To one girl, Eva, the story of matricide is the ultimate outrage, for her own mother died trying to protect her from this place. On her knees before a master she cries, 'I beg you to respect my grief. Kill me.' As she is forced to eat ordure with a spoon the bombers are heard over the mansion. 'Amazing that she baulks at such a delicacy,' croons a master. Another disappears into the sideroom to masturbate – a recurrent reaction to the intolerable excitement.

Following this repellent scene houserules are devised concerning the collection of faeces. Private defecation and disposal is forbidden. A tureen of ordure is served for dinner, over which shot Mussolini's voice is heard ranting about the glory of Fascist youth. Eva, breaking down, is told by another girl to 'say a prayer to the Virgin'. The madam resumes her tale of matricide.

A bottoms beauty contest follows. The winner will be executed, announce the masters. As they inspect the naked bodies they discuss the delights of anal perversion. 'The act of sodomy is special. It can be repeated indefinitely.' The winner has a pistol to his head. A blank is fired. 'You fool,' says his executioner. 'How could you think we wanted to kill you? We want to kill you a thousand times.'

There follows 'the circle of blood', in which the humiliations intensify into the final scenes of torture and murder. The sequence begins on the masters primping in drag, adjusting earrings and wigs to the background of bombers. In the Hall the victims wait. A master shouts, 'Clap and sing. We'll make you snivel for the rest of your days. The few you've got left.' Two of the madams prattle in French about the difficulties of having no money. 'Work with your hands,' they simper. 'Write any old thing.' Odd lines, these, signalling Pasolini's vision of late twentieth-century Europe – swelling armies of self-employed and a shrinking community of increasingly nihilistic artists. The madams themselves, of course, are adept at 'working with their hands' and 'writing any old thing', though never on paper.

An accordion plays. Meagre rituals begin, with the Priest playing Beelzebub in an infantile inversion of a religious ceremony. The masters take youths as their brides. Rings are exchanged. One 'bride', a shifty youth named Luigi, mutters, 'My friends are always ready.' Now follows the most degrading scene in the movie. A youth informs on a girl named Graziella, pointing out that she has a photograph of her boyfriend under her pillow. The informer triggers a chain reaction of denunciations. Graziella, terrified, announces that Eva and another girl are lovers. Eva, caught *in flagrante delicto*, tells the masters that one youth, Enzio, makes love with a black servant girl. Enzio is caught in the act, and barely has time to offer the Fascist salute before he is shot by masters outraged at this evidence of that ultimate obscenity, human affection. The black girl is shot at close range, as an afterthought.

In this way, the victims have been turned against each other. One squad of youths machine-guns 'the queens' (those youths misguided enough to remain heterosexual). An announcement is made. Soon the *ménage* must move operations from this mansion (near Mantua) into Salò. In the same way, Pasolini had to move back into the Cine Città studios to shoot the final scenes of his movie, those involving explicit acts of torture.

The closing sequence (heavily cut in the UK print) shows the escalation of the violence. The stories of the madams become wildly sadistic, while girls are mutilated in the yard by death squads of youths. The masters watch in turn from a window, using opera glasses.

Pasolini inserts two shots into this final circle of hell which serve to show his abhorrence of the spectacle. The pianist suddenly stops playing, walks to her room and throws herself from the window. Bombs thud in the distance. A girl is dragged to a stake; a lighted candle is held to her genitals. The masters jig as a youth is flogged to death and another is branded. But, counterpoising the shock effect of the pianist's suicide is the final shot, showing two youths dancing together indoors to the radio swing music. They are wearing their jackboots but they put down their guns to dance. One asks, 'What's your girl's name?' 'Marguerita,' says the other. Pasolini freezes the action at this moment, turning the two youths into a tableau, framed between the machine-like Futurist paintings which hang on the wall behind them. There is wanton cruelty in these two youths, but also the first stir of a return to reality – girls' names and dance music signal the imminent collapse of this short-lived 'inferno'.

What Pasolini adds to de Sade, apart from the major inspiration of using Salò as his context for the orgy, is the element of tenderness. Faced with verbal and physical violence, the victims are either subverted to the tyrants' cause or else cling to the residues of their former values of family. 'We Fascists are the only true anarchists,' rant the masters. 'We must submit our fascinations to a single action.' This histrionic idiocy, linked to sexual inadequacy and an obsession with political supremacy, *is* Fascism, Pasolini is saying, in its final doomed phase. The final scenes in *Salò*, in fact, are reminiscent of Visconti's *The Damned*, in which sexual perversion takes over as Hitler's regime falls apart in Germany.

What is curious about *Salò* is that implicit in its antipathy towards authoritarianism is a muted respect for family, that bourgeois institution which Pasolini, in earlier films, has criticized – notably, in *Theorem* and *Pigsty*. In times of nightmarish atrocity, family values represent the only form of coherence and human dignity available, Pasolini suggests in a number of brief moments. No other consolation but the memory of family love, and family sacrifice, is left to these shattered victims of the masters. A boy is marched off; his mother risks her life running after him with his scarf. Two girls cling together sobbing. In their distress, they have become sisters. The shattering of family bonds only breeds in certain of the victims a strong urge to reconstitute where possible those healing fragments they remember. Pasolini has never before shown us such a range of degradations (though *Pigsty* comes close to it) but neither has he shown us so clearly the *necessity* of tenderness. It was perhaps this new firming of capacities in Pasolini – his compression of intellect aligned with an emotional openness – which made him seem dangerous to the Italian public, and to some of the critics. *Salò* shows us how precise Pasolini was becoming as an artist. He knows exactly what he is attacking – the perversions of power politics, whatever the label, and the exploitation of the poor. He also knows exactly how he wants to formulate his attack. Gideon Bachmann, who watched Pasolini filming *Salò* near Mantua in 1975, was himself struck by the new urgency of purpose about the *auteur*, commenting in his journal of the film, 'I have always felt that Pasolini was tremendously hurried while shooting, and that he seemed impatient with the machinery ... this time ... I see at once that he is making actors repeat lines and actions in front of the cameras until he gets what he wants.'

Salò represents Pasolini's emergence from his domain of schema, dream and Gramscian theory into a world all his own, thoroughly synthesized and made over to his own artistic purposes, while remaining subversive. Tuned this finely, Pasolini, with his very definite political assertions and his immense cultural confidence, was indeed an awesome figure. It is hardly surprising that in Italy *Salò* was banned, and in London, during its first public showing in August 1978, it was seized by members of the Vice Squad (it re-opened with twelve minutes of cuts eighteen months later).

Pasolini himself was clearly aware of the artistic development *Salò* measured for he remarked that it was his first 'modern' movie. He was becoming aware of himself as a kind of psychotherapist in Italian society whose role was to exhume the material of trauma and confront a population with its own repressed fears and angers. 'I want us to realize,' he told Gideon Bachmann during shooting, 'that there are basic human instincts that must be recognized ... today we have come full circle, because what is being exploited is man's mind *and* his body. In consumer society we are being given a false sense of freedom, because we are suddenly allowed to

The Hall of Orgies in *Salò*

do things that had been taboo.' He refers to the now dominant notion that we 'must fight for equality in what we buy; we must all become, as in the business world, more cruel in order to succeed. Isn't that what Hitler wanted? de Sade was a romantic. He thought he was describing something special. Today we know he wasn't.'

Pasolini was an artist whose life tended to fulfil Wilde's dictum – he often put his talent into his work and his genius into living. While the notion of separating life from art is a little factitious, it remains to be said that Pasolini's intensity was of that electric, self-consuming variety which was difficult to condense in any one artistic format, be it literary, theatrical or cinematic. With the exception of the Trilogy of Life and *Salò*, his films always seemed a little less complex than the man. As early as 1968, he commented, in an interview with Lina Peroni (*Inquadrature* Nos 15–16, cited in Oswald Stack's Pasolini monograph), 'The sort of indignation and anger against the classical bourgeoisie no longer has any rationale behind it because the bourgeoisie is undergoing a revolutionary change: it is assimilating everybody.'

What, then, did Pasolini propose as a means of resistance to the spreading classlessness of greed which he saw afflicting all? 'In my language, "shut the schools, shut TV" means change,' he told Furio Colombo, in the last interview he gave before his death (*Tutti Libri* No 28). 'However, change so desperate and drastic replicates the situation ... I continue to assert we are all in danger.'

What was this danger? Perhaps it is the condition Bachmann describes, in his *Salò* journal, echoing Pasolini's own penchant for hyperbole, as 'the descent into the greatest conformism in human history'. That is, the retreat from authentic forms of dissidence which now marks Western intellectuals. 'Never as much as today has it been necessary for political men and intellectuals to be so anti-conformist, clear and hard,' Pasolini commented (*Screen* 49, July 1976). His tragic death on Ostia beach in 1975 left his career incomplete, yet far from obliterated. In his films, novels, poems, linguistic and philosophical fragments and his provocative articles for *Corriere della Sera* he has left us his provoking presence.

Finally, Fellini's *La Città delle donne* (*City of Women*), of 1980, in which he resumes his perennial theme, the futility of sexual relations, using the scenario-as-dream to create another autobiographical confession concerning the pain of self. The hero, Snaporaz (Marcello Mastroianni), is a later model of Fellini's previous autobiographical personae, Guido (*8½*) and Giorgio (*Juliet of the Spirits*). Like them, he is haunted by the Ideal Woman, but he

Pier Paolo Pasolini as a cleric in *The Canterbury Tales*

Anna Prucnal in *City of Women*

is also older. If his situation has not altered, society has. He is still ensnared by the vulpine female (Bernice Stegers) and the 'pain-in-the-ass' wife (Anna Prucnal), as Fellini describes the type. What has changed is the world of women around him. The girl he gropes on the train, the feminist militants he meets at the convention, the drugged punk road-waifs who give him a lift, the incorrigible guerrilla girl (Donatella Damiani), all live without men as a reference point. For Snaporaz, this proves intimidating. But for Katzone, the fading philanderer in whose erotic mausoleum Snaporaz seeks refuge, the rise of Liberated Woman is the death knell of everything that makes life worthwhile – to wit, seduction. Feminism, a force of modern life which Fellini satirizes in this movie, is as 'vital as water ... filling every available space'. British reviewers, however, detected a malicious caricature of feminist goals in his portrayal:

'By taking it out of context, by suggesting that it is a savage game acted out mainly by drug-oriented nymphomaniacs or lesbians ... Fellini pushes the whole idea towards farcical parody, grieved Alan Brien (*Sunday Times*, 20th September 1981). He perceived the film as 'the final capitulation of bogus virility faced by the reality of female sexuality'. Fellini, however, confronts the phenomenon obliquely. A key line occurs when the Girl from the Train sums up the feminist convention: 'We have been deceived. We've performed our rites. The eyes of man are upon us.' Feminism, Fellini perceives, is a *reactive* process, not yet self-defining. Opposition to *macho* goals is not in itself a solution, only a holding programme. Thus Fellini can be said to have 'a reserved attitude' towards feminism, as his former collaborator Lina Wertmuller has expressed it. He is equally satirical about the frailties of his two male protagonists, we might bear in mind: Snaporaz and Katzone represent the passive and active wings of the *macho* operation. Snaporaz is chained to his bitter wife, who shrieks, 'I'm not your refuge' (an insight Juliet arrived at with *her* errant husband Giorgio); while Katzone, his erotic temple commandeered by the feminist collective, his favourite guard dog shot on his doorstep, retreats into mother worship, kissing his late mother's bust and murmuring 'it is you alone I love'.

Bogus virility, certainly, is seen to capitulate. But what of authentic masculinity? Where, if anywhere, does this manifest itself? Fellini insists that he does not intend to travesty feminism – 'I have not invented a single word' – and, indeed, everything women in his movie aver, from the importance of yoga to the exaltation of the vagina, has been said by women. But there is a case for arguing that he caricatures masculinity, reducing it to sexual

Marcello Mastroianni and friends in *City of Women*

rapacity, madonna worship and emasculation by shrewish wives. The most significant limitation of *City of Women*, in short, is the absence of any *young* men in the movie (apart from the garish gays who flit through the interrogation scenes). How do young men relate to a world where feminism is the vital force? The young women we see are drugged, violent and euphoric. They shoot pistols at passing jets. One murmurs to Snaporaz from her dope haze, 'why not dance with us? There's nothing better to do.' Dance as despair not celebration is Fellini's bleak verdict on modern Italy, a country where women embrace feminist ideas more stridently than they do elsewhere in the West and where civic harmony seems absolutely out of reach, whatever the slogans are.

Like Pasolini, Fellini appears to have reached the conclusion, in his cinema, that sex is a suicidal form of self-deception; this motif underwrites *City of Women* more powerfully than that of the imputed malice towards feminists. 'A house without a woman is like a sea without a siren,' shrieks one old crone as she fusses over Snaporaz (only the crones in this movie do any fussing over men). It sounds like the end of feminine tenderness, perhaps, but it might be more equivocal. A sea without sirens, after all, sounds reasonably agreeable. As for a city without women ... ?

As usual, Fellini prevaricates. He is no partisan when it comes to social issues and current dogmas, merely 'a tourist'. His absence of hope in any solution, feminist or otherwise, seems appropriate enough, given his context of modern Italian cities. Where his film can be faulted, however, is in its stylistic slackness. Unlike Pasolini, who broke new ground in *Salò*, achieving great tension between his blankly glossy presentation and the horrors of the subject, Fellini shows some over-dependence on his staple imagery, the great fairground of childhood: Snaporaz on the helter skelter, Guido in the circus ring; Juliet in the biplane, Snaporaz in the air balloon. As for the film-as-dream, the sub-Borgesian time warp of the train heading (twice) into the tunnel, this is a retreat from reality, a sham ending which demeans the film, reducing it to a montage of *déjà vus*.

5. Debate and Denunciation: Francesco Rosi and Elio Petrie

'There is something visceral about power,' Francesco Rosi commented in 1976 (*Le Dossier Rosi*, Stock, Paris), 'something religious which has to be a faith ... there is a moment when, in opposing power, people cannot avoid being possessed by and participating in power. It is this dimension which one cannot analyse.' It is precisely this dimension, however, which Rosi and Petrie have tried to analyse in their feature films over the past two decades. The relation of the individual to authority is their common theme; relationships which exist on the basis of friendship or love are secondary to their intentions. In these two film-makers we observe Italian cinema at its most patriarchal.

Rosi first worked as assistant director with Visconti on *La Terra Trema*, and his association with Visconti continued through the Fifties – he was co-writer on *Bellissima* and assistant director/co-writer on *Senso*. His debut movie was *La Sfida* (*The Challenge*) in 1958. The film deals with corruption in the vegetable markets of Naples, Rosi's home city, and the attempts of a black marketeer, Vito, to by-pass the big rackets. He is shot dead by the leading mobster at the end of the film. Most of Rosi's eleven films contain gun murders. The theme of Neapolitan extortion is one to which Rosi returns, in *Le Mane sulla Città* (*Hands Across the City*) in 1965, and *Lucky Luciano*, in 1973.

In Rosi's second film, *I Magliari* (*The Rag Trade Swindlers*), 1959, Visconti's influence can be discerned, not so much in the theme, which concerns the Neapolitan carpet dealers who trade in Germany, but in the somewhat operatic love affair which occurs between an innocent salesman, Mario, and the mistress of a rival gangleader, Paula (played by Belinda Lee, who died not long afterwards). *I Magliari* is the only Rosi film (with the exception of his divertissement, *C'Era Una Volta*) to make an erotic relationship between a man and a woman central to its dramatic purposes.

Rosi began to attract wider interest with *Salvatore Giuliano*, made in 1961 with some illustrious collaborators – on the script Susi Cecchi D'Amico (Visconti's close friend and long-term collaborator), Enzo Provenzale and

Francesco Rosi during the shooting of *Christ Stopped at Eboli*

Franco Solinas are all credited, besides Rosi, while the cinematography was undertaken by Gianni Di Venenzo. Described by P. J. Dyer (in *Monthly Film Bulletin*, June 1963) as a 'complex political and social case history of the Mafia of post-war Sicily' the film is pivotted round the death of a Sicilian bandit, the Giuliano of the title, who in fact only appears as a bullet-shattered corpse in the opening scene. Much of the film takes place in a court-room at Viterbo, where Giuliano's fellow bandits are arraigned; flash-backs showing the massacre of communist workers at a rally in Portella della Ginestra and Giuliano's betrayal by his henchman Pisciotta are inserted into these lengthy debating scenes.

Giuliano, for all its quasi-documentary effect, is actually a carefully contrived study of violence, owing much to Visconti's method in *La Terra Trema*, in which Visconti used Sicilian fishermen as Rosi here uses the mountain peasants. Rosi's handling of the scenes outside the court-room may look neo-realist but his attention to the litigious aspects of the affair is strictly his own enthusiasm. Rosi had studied Law as a young man.

In 1961 Elio Petrie made his debut feature, *L'Assassino* (*The Assassin*), a *film noir* shot in the then modish manner of Antonioni. Up until this time, Petrie had been working first as a journalist, and then as a scriptwriter for Giuseppe De Santis.

The Assassin concerns an antique dealer named Poletti (Marcello Mastroianni), who is arrested by the police and charged with the murder of his mistress. The film shows Poletti's breakdown in the cell; he torments himself with confessions of past misdeeds, appearing to forget that he hasn't committed the crime. Poletti's problem, as Petrie sees it, is his blistered conscience – 'he is innocent of murder but guilty of inhumanity.' Elements which recur in Petrie's cinema are present in this movie. The indifference of the police, who snooze during interrogation, and the oppression of women, whom Petrie delineates repeatedly as 'at the base of the social pyramid', as he puts it.

Reviewers in the UK have tended to like *The Assassin*. Peter J. Dyer described it, in *Sight and Sound* (Winter 1961/2), as 'nonchalant, subliminal, drenched in a nightmarish brew of farce and intangible menace', while much later John Coleman in the *New Statesman* (28th May 1971) referred to it as 'the best of those Petrie films I have seen'. Petrie's second feature film, *I Giorni Contati* (*The Numbered Days*), made in 1962 (and starring Salvo Randone, an actor Petrie has used consistently), is a film which seems consistent with his general political intentions.

Randone plays Cesare, an elderly plumber whose life suddenly undergoes a shift of values when he sees a man of about his own age die on a tram, presumably of a heart attack. Cesare gives up his job and drifts about the city (Rome), visiting a painter he knows, making a visit to the country, striking up a conversation with a beggar, and trying to make contact with his daughter.

The film, while it bears some superficial resemblance to De Sica's *Umberto D* and Kurosawa's *Ikiru*, is autobiographical in some of its detail: Petrie's father was an artisan who, at around the age of 50, made a similar decision to 'rest'; while the somewhat surprising forays of Cesare into the studio of a painter reflect Petrie's fascination with the painter's terrain – he was involved soon after this in making a short 16 mm. documentary film about the American Op painter, Jim Dine, which in its turn exerted a direct influence on his 1968 feature movie about a painter, *A Quiet Place in the Country*.

Petrie has said that *The Numbered Days* is a film 'against work'. In this sense, the film can be related to his later feature, *The Working Class Go to Heaven*, which depicts the world of a factory engineer. Petrie, however, insists that an artisan, like Cesare, is to some extent self-governing, unlike the severely constricted factory worker, Lulu, in the later film. In 1963 he adapted a novel by Lucio Mastronardi, *Il Maestro di Vigevano*, for the screen, and his producer, Dino de Laurentiis, decided that the film must be a vehicle for the comic talents of Alberto Sordi. In the event, the resulting film, based on a new screenplay by the comedy screenwriters Age and Scarpelli, was in Petrie's words 'completely mistaken ... it contained too many contradictory elements'.

There followed two more bread and butter films, the sketch *Peccato nel Pomeriggio* (*Sin in the Afternoon*), one of several assembled under the portmanteau title of *High Infidelity*, and then, in 1966, his future shock thriller *La Decima Vittima* (*The Tenth Victim*), based on a thriller by Robert Sheckley.

The Tenth Victim, shot in Rome in 1966, concerns a computerized manhunt which is televised for the entertainment of the sensation-hungry nation around the year 2000. Killing has now become a State-approved activity, and this is the tenth international manhunt. The hunter selected by the computer is a suave philanderer known as 'The Baron' (Mastroianni). But this time a second computer selection has been made; the Baron is now cast as 'the victim'; his killer will be Caroline Meredith, also a veteran of previous manhunts (Ursula Andress).

Using this lurid storyline, Petrie manages to inject some pointed satire, not least against Italians. Moon worship, hormone shots and various mid-Sixties' fads seam the satire, until the killing takes place during the shooting of a TV commercial. Confronted with this *fait accompli*, the Baron remarks gracefully, 'What a stupendous death you have rigged up for me. I am enchanted with your organization.' In the event, he does not die, and the last part of the film becomes garbled. Petrie's films often reveal this tendency to drift in the final stages.

At this time, while Petrie was occupied with bread and butter films, Rosi was consolidating the impact generated by *Salvatore Giuliano*. In 1963, he made *Hands Across the City*, starring Rod Steiger as a corrupt property developer, Nottola, who is running in the imminent city council elections in Naples, and suddenly finds himself the centre of a seething political controversy when a building belonging to one of his companies collapses.

Appeals to De Vita, a Left councillor, the only character to show any conscience, are not much help, the outraged citizens discover. 'You voted them in,' he argues, referring to the corrupt councillors who have been bribed and otherwise manipulated by Nottola. De Vita also tells Nottola, 'Flats should be built where the law permits and not as you wish. You're an outlaw.' Nottola offers money to the ruling Christian Democrat faction in return for the post of housing commissioner, should he be elected. He is warned that his commitment to the CD cause will cost the party votes. He is unpopular following the collapse of the tenement, in which two have died.

The elections duly take place, amid scenes of public riots, Nottola's foxy face gazing down from countless posters. The new mayor, De Angelis, crushes the Right with an intimidating strategy, forcing a public reconciliation between Nottola and his former political adversary, a CD member named Maglione.

De Vita remains aloof from these machinations, content to make a formal denunciation of Nottola before the courts, in which he alleges that money from a special development fund has been diverted. The court scenes are of special fascination to Rosi, who has remarked that he uses them when possible in his movies as 'a sort of remorse for not having finished my legal studies' (*Film Quarterly*, Spring 1965). The film ends with no conclusive case against Nottola established, for Rosi does not attempt to provide any kind of dramatic gloss to the reality he depicts. He says, 'I try to analyse that which exists around me, to find a story in it ... to relate a condition, to tell a city.'

The film ends with the inauguration of a vast public works site. We close on Nottola standing by a building as a piledriver starts up, with helicopter shots drawing us away from the vanquished city of Naples.

G. Nowell-Smith, considering the impact of the film in *Sight and Sound*, points out that the film has an epic structure which underlies its apparently naturalistic surface. In fact, Rosi's cinema depends very much on this epic aspect. In his next two films, this becomes more apparent.

Il Momento della Verità (*The Moment of Truth*), of 1964, was shot in Spain and written by the Spanish writers Ricardo Suay and Pedro Portabella. With the exception of a small cameo role played by Linda Christian, all the main roles were played by Spanish actors. The young hero Miguel dreams of becoming an El Cordobes, who receives a fortune for each fight: instead, he has to enrol in a tacky bullfight gym, where he is taught by a pompous old matador that a bull is *'una cosa sacra'*. The mechanics of training are cannily illustrated – 'They all want to fight but when the time comes they're wet with fear,' observes one ringside cynic, and it is a remark which is repeated twelve years later in Rosi's cinema, only this time from the mouth of a political pundit, observing a bunch of chic radicals at a party (*Cadaveri Eccellenti*).

In *Moment of Truth* Rosi's affinities with Visconti are again apparent. Miguel the matador, like Simone the boxer (in *Rocco and His Brothers*), suffers the same brief moment of success, is lured down the wrong road by a handsome woman, and ends up ruined. Simone commits a murder; Miguel is fatally gored in a contest for which he has no more relish. His last thoughts are of his mother, who will have seen him gored on television. The film ends on a funeral procession as oppressively baroque as the opening sequence outside the cathedral.

There followed a hiatus in Rosi's career until 1967, when he prepared with Tonino Guerra an adaptation of the Neapolitan writer Basile's fairy tales. In its original version it would have represented for Rosi what, maybe, *Satyricon* represented for Fellini two years later – an excursion into mythic lands. In the event, Carlo Ponti agreed to produce only on condition that Guerra/Rosi modify the script as a vehicle for his wife, Sophia Loren. While the film reveals a surprising talent in Rosi for scenes of earthy comedy (with Sophia Loren at her most fulsome), *C'Era Una Volta* (*Once Upon A Time*) is hardly the film that Guerra and Rosi envisaged. As Guerra points out in an interview in Jean Gili's *Francesco Rosi: Cinéma et Pouvoir* (Editions du Cerf 1976): 'Many of the minor characters disappear in the final script ... along with the magical and fantasy elements.'

Towards the end of the Sixties Petrie made two films which he regards as his first truly personal films. *A Ciascuno il Suo* (*To Each his Due*) was based on a novel by Leonardo Sciascia, about 'the cultural autism' of the intellectual, as Petrie describes it. Professor Laurana is living with his mother and grandmother. His place of work is a mere refuge for him and so is his home: by 'autism' Petrie means to imply a kind of emotional stasis.

Laurana lives in Sicily, in a village called Cefalù, and becomes accidentally entangled in a Mafia murder. Laurana decides to confess the identity of the murderer, but in his naïvety he confesses to none other than the killer, and is murdered himself at the end, being thrown from a mob car.

Once again, Petrie has to accommodate a plot which stretches credibility. He himself justifies the ingenuousness of Laurana by pointing to the man's provinciality – 'Cefalù is not Palermo,' as he expresses it – and also in terms of Laurana's sexual neuroses – 'he complicates things, impelled by his complexes, his sexual frustrations.' As is usual with Petrie, his willingness to discuss the psychological disposition of his characters tends to create an eloquent fabric *outside* the film text itself.

The following year, 1968, Petrie made a film based on a script he had written in 1962, entitled *Un Tranquillo Posto di Campagna* (*A Quiet Day in the Country*). The film featured Franco Nero and Vanessa Redgrave, and concerns the inner world of an abstract painter, who is haunted during a period of rural idyll with his wife by the nightmarish images of a girl he once knew – who has since been murdered.

Petrie is as much interested in the painter's work as in his sexual turmoil, however, and makes use of twelve of Jim Dine's canvases. The film is notable for the remarkable photography of Luigi Kuveiller, who uses slow-motion footage to suggest paranormal events.

In 1969, shortly before he embarked on his most successful internationally known movie, *Investigation of a Citizen Above Suspicion*, Petrie contributed to a collective documentary about the mysterious death of Pinelli, the anarchist arrested by the police on suspicion of planting bombs who was later thrown from a police station window. The event was reported in the British Press by Mike Balfour, of the *Socialist Worker* (11th December 1971), who commented, 'It became clear in the subsequent court-case that Pinelli's "suicide" was nothing of the sort. The police accounts were full of contradictions and lies. His case became a symbol of injustice ... In 1971, in fact, the inquiry was re-opened, and the CIA-trained interrogator of Pinelli, Calabresi, was tried for manslaughter. At the time Petrie and Nelo Risi completed their documentary of events, however, the matter was still a political mystery. Only Petrie's twenty-minute section concerning the trial, and Nelo Risi's lengthier footage on Pinelli's background as an anarchist convenor, were ever completed.

Uomini Contro (*Just Another War*), which Rosi produced and directed in 1970, is his angriest film to date. The time is 1916, the place the front line of Carso, in the mountains above Trieste. Here an obscure Sardinian infantry regiment is fighting Austro–Hungarian troops. This theatre of war, less crucial militarily than the Western Front, exhibits a low level of morale as a consequence of its strategic insignificance. As one foot soldier puts it, 'what the hell are we fighting for – for pebbles?'

Mark Frechette faces the firing squad in *Just Another War*

Like Losey's *King and Country* and Kubrick's *Paths of Glory, Just Another War* makes the powerful point that the real enemy, as far as the soldiers themselves are concerned, is their own commanding officer, General Leone (played with unnerving fanaticism by Alain Cuny). He orders his men into battle against overwhelming odds, and orders his reluctant lieutenant, Ottolenghi (Gian Maria Volonte), to shoot a soldier following an attempted mutiny. Finally a mutinous lieutenant named Sassu (Mark Frechette) is executed by a firing squad.

Both Losey and Kubrick, centring their stories around a bewildered private soldier destroyed by an impersonal war machine, make capital out of the inhumanity of militarism, using the device of court martial scenes. Rosi, on the other hand, is concerned, as he usually is, with analysing the underlying struggle between foot soldier and officer. He treats the two junior officers, Ottolenghi and Sassu, as crucial pivots in the changing consciousness of the soldiers' resistance and mutiny. Ottolenghi, the older, more battle-weary of the two, is sick to death of battle.

Sassu, however, a new arrival to the front, has yet to lose his love of glory. In the trenches he and Ottolenghi argue. 'Would the situation be better with socialism?' Sassu asks dubiously. Ottolenghi replies, 'The only way is to get shooting and see.'

165

Gian Maria Volonté in *Investigation of a Citizen above Suspicion*

Where Rosi has been particularly successful, one feels, is in the character of Sassu, an interesting hero in Rosi's pantheon of fighting men, being genuinely dynamic, as opposed to the hypermanic Mattei and Nottola, or the sullenly passive Luciano. Sassu arrives, a quiet conformist, and dies a steely, dissident veteran. His death leaves the film open to the query, 'Which Italy is being fought for in this bloody confusion?' The question, as always in a Rosi film, remains unanswerable, but disturbing.

Petrie's film of 1969, *Indagine su un cittadino al di sopra di ogni sospetto* (*Investigation of a Citizen Above Suspicion*), also treats the authoritarian theme with considerable anger, though now the setting is Rome, and the time is the present. The film was co-written with Petrie's regular scenarist, Ugo Pirro, and again photographed (with great effect in the opening sequence) by Luigi Kuveiller. Petrie has said that he and Pirro wanted to make a film 'against the police'. In a collection of interviews and essays edited by Jean Gili (*Elio Petrie*, Faculté des Lettres et Sciences Humaines, Nice 1974), Petrie elaborates this theme: 'The police of the Republic of Italy, in the twenty-five years which have followed the end of fascism, in spite of the abolition of capital punishment, have committed in the streets and squares numberless summary arrests ... No policeman has ever been arraigned for any murder ... thus I was interested above all else in this movie to describe the mechanics of immunity before the law guaranteed to the protectors of the vested interests.'

Petrie's method of examination, however, is somewhat different from that employed by Rosi in his later film, *Cadaveri Eccellenti*, which also deals with the theme of corruption and homicide in high places. Petrie and co-scenarist Pirro treat their bleak theme satirically rather than dispassionately. The opening sequence sets the tone for what follows in bizarre fashion.

The anti-hero, the police chief (Gian Maria Volonte) of the Homicide Division, visits a girl he knows, Augusta, in her apartment. She is a part-time whore, a friend of student militants, a dope-smoker. 'How will you kill me this time?' she asks him. The chief replies, 'I'll cut your throat.' They make love, the camera above them. Suddenly Augusta arches – in ecstasy, or agony? Blue light bathes her back as she falls and blackness fills the frame. An accomplished visual requiem for the chief's victim, for he has kept his word and slashed her throat. He rises, wrapping himself in a white sheet, rinses blood from his hands, takes a drink, rests dreaming on a couch, looks through the window at Hebriac inscriptions on the walls of a synagogue opposite, and begins to 'dress' the room as a 'set' for a homicide, carefully extracting a ring from the dead girl's finger. He leaves (after hesitation) a pile of money he finds, but takes some gems. In the hall he phones police headquarters to announce that a murder has taken place, giving the victim's full name and her address, and signing off anonymously. As he leaves the apartment building he bumps into a young man entering. The look exchanged between them is quizzical, and deeply antagonistic.

Only now does it become plain that the killer is in fact the police chief. He

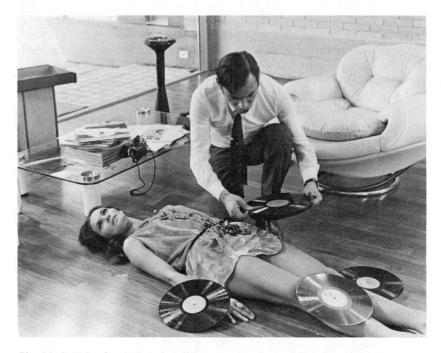

Gian Maria Volonté and Florinda Balkan in *Investigation of a Citizen above Suspicion*

returns to his office at headquarters. Deferential subordinates offer congratulations – he has been promoted. He opens champagne (filched from Augusta's flat) and 'learns' of the murder. Alone in his office, at his desk, under a wallmap of the city, he becomes the man of power. The shot, as he sinks into the chief's chair, matches his moment of triumph rising from the bloody body.

Now begins the homicide squad's investigation, in which innocents are arrested, a number of them political subversives already on record. The chief's hubris is intensified. 'The only guilty one is me,' he boasts, drawing nervous giggles from his juniors. Via a series of terse flashbacks, Petrie shows his past relationship with Augusta. She used to pose for 'crime' photos, playing mutilated 'corpses' for the chief's lubricious camera. 'A sexual kook' is the verdict of the chief's staff, once the flat has been searched. These flashbacks develop the sexual inadequacy of the police chief and Petrie draws links between this and his power hunger.

What Petrie shows us is how the depraved guile of the chief is matched by the ironic dissidence of a prime suspect, the student revolutionary named Pace, who is interrogated following a period of surveillance. Pace is the same young man the chief met as he left the girl's flat.

What frightens the chief about Pace, alone among his suspects, is that he cannot seem to intimidate him. Pace feels no more respect for the power of the police than the chief does.

It emerges during the interrogations that a number of these young men are second generation dissidents; their fathers and uncles are also on file for anti-fascist activities during Mussolini's era. Against this 'hallowed' tradition of anti-authoritarianism the chief can only rant, 'Revolution is syphilis.' He comes to understand that their attacks on the vested interests are not self-promoting, as his are, but iconoclastic, hence their absence of unity and hierarchy.

Pace, however, knows that the chief murdered Augusta but refuses to denounce him. Pace has power, but he refuses to wield it. In this subtle scene, Petrie makes the point that each citizen has a right to become an investigator, which implies not only increased wariness about the meaning of the social institutions, but also increased curiosity. As Petrie dryly expresses this paradox, 'I'm 43 and I've never understood what happens on the Stock Exchange.'

The chief is so disoriented by Pace's contempt that he writes out his own confession and goes to tell the commissioner, 'You'll never understand my gesture. I want to reaffirm the principle of authority. If you want to interrogate me I'll be at home.'

In his flat, the chief waits, packing his whisky and books, his police diplomas, his photographs of violent crimes, and reading Article 247 on house arrests. The commissioner and his staff arrive and the chief prepares himself for a ceremonial self-humiliation. The police officers, however, deny his guilt. 'You are insulting your colleagues,' the chief is warned. 'Where's your motive?' The chief replies, pathetically, 'She made fun of me, and the whole establishment. Day by day she exposed my inadequacy.' He goes on his knees and receives a paternal pat of ritual expiation.

This has been a dream sequence. Outside, the police arrive. As the officers file in, forming a frozen tableau, a quotation from Kafka is introduced – 'whatever impression he makes on us, he is a servant of the law, to which he belongs, and thus escapes from human judgement.'

Petrie sees *The Investigation of a Citizen* as the first film in a trilogy, which he completed in the Seventies when he made *The Working Class Goes to Heaven* in 1972 and *Property is No Longer a Theft* in 1973.

Working Class . . . centres around one worker at a Turin plant, Lulu Massa, an engineer in his late thirties (again played by Gian Maria Volonte), responsible for supervising equipment, piece rates and workers. The film, while made in the early Seventies, pertains to industrial conditions prevalent in Northern Italy during the mid-Sixties, at a time when union representation was only beginning to acquire mass impetus. Industrial unrest depicted in the film arises from the *'cotimo'* system of piece rates, which the skilled workers believe to be detrimental to their main interests. In the early sequences of the film it is Lulu's invidious duty to reinforce company rules regarding *'cotimo'* production methods.

Lulu has separated from his first wife, by whom he has a teenage son, Arturo. He is now living with another woman, by whom he has another child, and in the early domestic scenes we realize that this relationship, too, is cracking under mutual pressure (the woman has also been working for over fifteen years, in a shop).

A further axis beyond that of plant floor, picket gate and domestic crisis, is the asylum, to which a popular elderly worker named Militina (played by Salvo Randone, developing his role as the 'retired' steel worker in *I Giorni Contati*) has been relegated, following a breakdown. In the first of the two important scenes at the asylum, Militina tells Lulu he sees madness as the only escape the common man has.

Militina serves as the 'seer', pronouncing on the calamity of industrial life with something of Petrie's own anger. 'A man must know what he is doing,' he tells Lulu; it is perhaps the gravest line in the film, for no one does. In the second scene between Lulu and Militina, near the end of the film, Militina demonstrates that even here a measure of solidarity can emerge, when he beats on the walls of his cell and many others join him. Lulu, however, is cast as a mediator between workers and bosses. He can identify with neither group; this induces his own crackup.

The effectiveness of *Working Class* ... derives from Petrie's decision to isolate his hero from *both* the dominant power structures which surround him. He is a man who 'only partially understands what is happening', Petrie has declared (*Cinéaste* Vol 6 No 1). 'He is a slave to the bosses, a TV watcher, and anarchist, a member of the middle-class, a revolutionary, and a trade unionist, as we all are.' Factory work, Petrie is convinced, produces loss of libido. At one point in the film a worker complains that he cannot work at his bench standing up because every time he stands he wants to urinate. Petrie is quite explicit on this point when he comments (in the monograph *Elio Petrie* edited by Jean Gili), 'Treating our sexuality as an economic commodity is evidently a form of global precariousness and poverty of spirit.'

Rosi's *Il Caso Mattei* (*The Mattei Affair*), featuring Gian Maria Volonte, portrayed Enrico Mattei, President of Italy's state-owned oil company, ENI, who sought to make Italy self-sufficient in oil and gas supplies by exploiting the methane layers in the Po Valley and, when the yield failed to show, went shopping outside the existing cartels. In October 1972, returning from a good will tour of Sicily in his private aeroplane, Mattei was killed, along with his pilot and a journalist named McHale, when his plane mysteriously crashed coming into Rome airport.

Rosi and co-scenarist Tonino Guerra structure the film narrative in three intersecting sections. First, there are the scenes leading up to and including the plane crash, in which we see the tycoon isolated in his hotel rooms during the Sicilian trip, following threats upon his life. We also see Mattei in the plane with the journalist and pilot, watching the moon through a thunderstorm, speculating ironically on the oil possibilities there; and finally after the crash,

Gian Maria Volonte in *The Working Class Goes to Heaven*

in which the world news agencies buzz, eye witness accounts are garnered, pundits speculate about the political significance of Mattei's death, and a commission of inquiry is established.

Rosi also shows the rise to power of Mattei in the state oil industry after the end of the Second World War, liaising with a former AGIP (later ENI) engineer named Ferrari over a wartime report Ferrari wrote about oil in the Po Valley. Ferrari shows Mattei the sites. Mattei immediately understands

171

that large fields in the Po could help him realize his dream – to disconnect Italy from her dependence on the United States by developing an independent energy policy based on Po oil reserves. At much the same time AGIP becomes ENI, a conglomerate of State-owned and financed companies which Mattei, as President, comes to dominate. Mattei at the centre of political power is the main drive of these scenes, showing his capacity to make or break companies by offering loans or withholding them.

Rosi realizes his own purposes by pushing his film inquiry into the 'thickets of ambiguity' surrounding Mattei's death, as Philip Strick aptly described it. Did the plane explode in the air or on the ground? Was a bomb planted at the airport in Sicily while the airport police were inexplicably off duty? Were the Mafia behind the crash, or the American 'imperialists', or both? Why did Mauro, the journalist investigating Mattei's last days, suddenly vanish in 1970, shortly before Rosi and Guerra began to assemble the script for the film? So crucial does Rosi see the investigative role in this, his most densely documented movie to date, that he appears himself in one or two of the Sicilian scenes, in the persona of Rosi/Mauro, asking awkward questions of the airport barman, and getting nowhere.

Rosi's detached debating procedure enables us to see 'round' Mattei's public image, so to speak, and glimpse the vulnerable aspects, not least his defensive attitude towards American oil tycoons, whose rebuffs are largely responsible for triggering Mattei's global oil-shopping; he visits Iran and Moscow to clinch deals on behalf of AGIP; and he defends his intervention in North African oil markets on the grounds that the Seven Sisters will not admit Italy into the monopoly. Meanwhile, AGIP/ENI is exposed in a TV documentary for accumulating large debts. Mattei reacts angrily, demanding company cutbacks. During a global tour, accompanied by a journalist, he attempts to defend his contributions to the neo-Fascist organizations. Mattei is unrepentant – 'I use them like taxis,' he says. 'I get in and pay. I refuse to give in to inactivity, like most Italians.' The global tour moves into the Indian ocean, to an Italian-built oil-rig. 'Oil makes governments fall, creates world balance,' Mattei argues, brushing aside a sneaky question about his alleged budget for callgirls, as he earlier ducked a question about his use of a private plane when ENI executives no longer have chauffeur-driven cars. Thus, Rosi shows that, latterly in his career, the gap between what Mattei is and what his public image drives him to become is narrowing dangerously. 'They have to kill him, he's too dangerous for Italy,' comments one reporter shortly before the plane crash.

Philip Strick, reviewing *The Mattei Affair* in *Sight and Sound* (Summer 1975), has remarked, *à propos* of Rosi's movies in general.' 'It is Rosi's selection of the available facts which ... we have to take on trust ... if Rosi has done all the research without reaching a verdict, how can we be in the position to draw any conclusions?' The answer, perhaps, is that to try to draw specific conclusions is to approach Rosi's kind of cinema with the wrong expectations. Rosi's value is, of course, his relative dispassion. As

Gian Maria Volonte in *The Mattei Affair*

Guerra observes, it is the impartiality of Rosi's techniques which lends his films their durability, long after more stridently committed movies have withered – 'they mellow like fruit,' he says, appositely.

Rosi himself has said of *The Mattei Affair* (*Cinéaste*, Autumn 1975) that he sees the inquiry as far from concluded. Interestingly, in this context, Anthony Sampson, writing about Mattei in 1975, in his book *The Seven Sisters* (Hodder and Stoughton), notes that, 'Mattei has been a mythical figure to the Italian Left, and rumours have circulated that Mattei's plane had been sabotaged by one of his many enemies. But such a drastic reprisal was not at that stage called for. Mattei had already prepared to make peace, and his company's financial base – as soon emerged after his death – could not support the fight much longer.' While Sampson cuts down the assassination myth, however, he does concede that Mattei's role in world oil politics was decisive: 'He had shown the producing countries that the cartel could be broken. He had forced the Seven Sisters to cut the profits of the producer ... and he had encouraged other consuming nations, too, to form their own national companies in rivalry with the majors.' It is to the credit of Rosi's movie, indeed, that all these significant elements to Mattei's career were clearly isolated in his film inquiry three years earlier.

In the third film of Petrie's trilogy concerning social schizophrenia, *Property is No Longer a Theft*, there is a curiously uneasy balance between the narrative and the use of expressionist devices – monologues to camera and montage sequences. Petrie insists that, like Brecht's theatre productions, his cinema, though stylized, has a popular base. But does it? What is most apparent about his political trilogy is that the emotional impact diminishes from one to the next. The most inaccessible of the three is *Property is No Longer a Theft*.

Petrie takes as his raw materials for a debate on money a well-to-do butcher in Rome and his tormented relationship with a young bank clerk who disrupts his materialist ambitions. The clerk first sees the butcher meeting the bank manager to ask for a large loan, in order to expand his burgeoning slaughterhouse operation. The butcher gets his loan. When the clerk requests a small loan from the manager, however, he is rejected. The clerk burns a ten thousand lire note in front of the manager as a gesture of protest and resigns his post.

In the scenes which follow the clerk begins a campaign of harassment against the butcher. This process involves not only the butcher, his girl-friend Anita, and himself, but also his father (an ex-actor) in a complicated *danse macabre* of enmity and strange attraction.

It is clear, then, that Petrie (and Ugo Pirri) have approached the difficult theme by trying to schematize the material. The characters are types, and vent their frustrations in the form of monologues. The butcher says, 'Capital enables me to live. I get rich through the people who have nothing.' Anita, however, in her monologues declares: 'I am many pieces ... I laugh because you like me as you like mineral water.' Between these two poles, the exploiter and the exploited, the clerk equivocates.

Money is a cancer, for Petrie, as revolution was a 'syphilis' for the chief in *Investigation of a Citizen* ... 'Everything to do with money can become dangerous,' Petrie says; 'violence, aggression stem from it. A bank is a forti-fied church.' But what Petrie seems to lose sight of is that a good deal of what he is saying is already obvious.

Petrie, then, fails to establish any form of *vital complexity* for his debate to thrive on. The curious contradictions of money – that it makes men greedy but also sometimes liberates in them a selfless or generous impulse – is not considered. As a film about the exploitation of those who suddenly plummet into a money world from nowhere, it seems less coherent than, say, Fassbinder's *Fox*, which shows with some precision what happens to a poor boy from the Munich back-streets who suddenly wins a fortune on the lottery. Petrie's mistake, perhaps, is in failing to contain his ideas about the evils of money within a stable narrative framework.

In the same year, Rosi realized his film on *Lucky Luciano*, which has been described by Norman Mailer as the most intelligent film ever to treat the theme of the Mafia. For his leading actor, again Gian Maria Volonte, this role was a peculiar challenge, not only because he resembles the mobster (during research an interviewee who had known Luciano mistook Volonte for the late gangster) but also because Rosi treats the character of Luciano in a restrained, unexpressive vein.

'I took the character of Luciano as a vehicle for discussion about the Mafia,'

Rosi has stated (in *Le Dossier Rosi*, Stock, Paris, 1977). 'Why a gangster? Why Luciano? Because he provides a key to understanding the relationship between legal and illegal power, and the way they interconnect. Luciano was the first criminal genius to understand how to put illegal power at the service of legal power.'

The difficulty Rosi has to overcome in this study is how to convey the 'genius' of Luciano, given that he chooses to eschew the 'charismatic' approach towards the characterization. Luciano, returning to Naples after serving nine years of his American prison sentence, sets about restoring his links with the criminal leaders. Rosi shows his involvement in drug-smuggling links with Marseilles and New York, chiefly through the filmed accusations made against Luciano by the US Narcotics Bureau investigator, Charles Siragusa, who plays himself in the inquiry. But he fails to produce a complex character. Rosi has previously shown a tendency to blur the psychological outlines of his heroes and convert them into emblems for factual inquiries. In *Salvatore Giuliano*, as noted, we never see Giuliano alive. Something similar is true of *Illustrious Corpses*, in which the hidden hero of the drama is not so much Inspector Rogas, the man who investigates the multiple murders, but the enigmatic pharmacist, Cres.

Rosi, however, justifies his approach by pointing out that he is concerned with the issues raised by a man like Luciano, rather than the man himself. With this in mind, we should consider what issues are provoked by Luciano's manoeuvrings. The first, as is usual in a Rosi film, is the issue of complicity. Luciano is freed by Dewey in order to help establish an 'entente' in Southern Italy between the American liberating forces and Sicilian *mafiosi*, whose goodwill is needed in order to obtain supplies of raw and engineering materials. Commenting on this issue, Rosi has said, 'I began to think that the Mafia today represents a guarantee of the legitimacy of the system.'

This idea is one of plausible repellence. Those channels through which a reassertion of the morality of our institutions might be made, such as the United Nations, are shown in this film to be well aware of the problem (in this case, Luciano's involvement in heroin smuggling) but powerless to act usefully.

The film begins with Luciano's send-off at the docks, following the caption, 'In 1945 the USA made a gift to the Mafia ...' This leads into a flashback – Scarpano's restaurant in 1931, and the killing of a mobster while Luciano, excusing himself from the poker game at which the gangster dies, reappears afterwards to announce of the victim, 'He was like my own father to me.' Back to the docks: among the mobsters who come to wave goodbye are Albert Anastasia, Frank Collio and Lansky. 'Christ, what a party,' grates Siragusa from the police car. We next see Luciano's career as a rising racketeer via telescoped flashbacks of massacres, police bribes and criminal banquets, all of which show Luciano as 'the boss of bosses'. The chilling edict is pronounced, 'He who puts things in order is a man of power.' This

Gian Maria Volonte in *Lucky Luciano*

proposition, always the nervous system of a Rosi movie, seems to recede in importance after Luciano's arrival in Naples; initially, he appears to be making a sentimental journey, wandering through the churchyards checking tombstone inscriptions. From these scenes, with their ironically elegiac note, we move forward to 1952, and a United Nations conference on drug smuggling, during which Luciano is romantically described as 'a calm man with sad eyes who lives in Naples' (poetic understatement *par excellence*) and the Italian delegate gets defensive about Italy's key role in global trafficking, protesting that heroin is supplied from 'respectable pharmaceutical sources' and adding that 'it can be very useful'. The delegate also makes the more salient point that hundreds of undesirable aliens besides Luciano have been shipped home from the USA, following the American landing in Sicily.

Asked why he avoided presenting Luciano as directly involved in Neapolitan and Sicilian racketeering, Rosi commented with his characteristic prudence, 'A political film must not only be honest, but also precise.' Certainly, Siragusa's reports for the Narcotics Bureau and the Kefauver committee, which are quoted verbatim in the dialogue, never established substantial proof for Luciano's involvement in the heroin and prostitution rackets.

What Rosi does manage to show, however, is how Luciano reinserts himself in the Southern Italian power struggle. He holds a press conference, for example, and liaises with the ebullient New York mobster, Giannini (Rod

Steiger). During their initial meeting they pace the ruins at Pompeii together. Luciano makes it clear he doesn't expect to be hustled hard – 'My friends visit me for pleasure. They come as tourists.' This is a deft line (which bears the smack, one suspects, of Tonino Guerra's hand in the script – Guerra was only used at a late stage of script preparation in this instance, the basic work being done by Rosi himself and Janussi). It is later suggested that Luciano may have had a hand in Giannini's subsequent death; on his return to New York Giannini is gunned down in the street before he can testify against Luciano at one of Siragusa's Narcotics Bureau tribunals.

Towards the end of his life, Luciano oscillates between the desire to repress the past and the desire to legitimize his reputation. He donates cheques to the Church, and agrees to help an American writer who is preparing a screenplay about him. At the same time, Rosi shows the official pressure against him building. His house is searched. He is called for police questioning, tries to justify his mysterious acquisition of land and business operations in Palermo, and suffers his first heart attack, while under constant surveillance. At the airport, where he is due to meet the American writer, he sustains a further, fatal heart attack. The film ends by emphasizing the circular mechanics of power, the verdict coming from a reporter talking to Siragusa, 'You keep chasing Luciano, we keep chasing Kefauver, and we all find ourselves at the same place we started.' Over the shots of Luciano's corpse, up comes the music of Glenn Miller.

In 1976 Rosi made *Cadaveri Eccellenti* and Petrie *Todo Modo*. Both films were adapted from novels by Leonardo Sciascia, the Sicilian author who at this time became a Communist Party deputy, following Berlinguer's directive that, when possible, intellectuals of the left should be drawn directly into the PCI. *Todo Modo* takes Petrie further from his intention – to make films with which popular audiences can identify. A group of well-known Christian Democrats closet themselves in a remote retreat ('a spiritual gymnasium,' as Petrie has ironically described it) to revise their attitudes and aims *vis à vis* the Catholic faith.

Petrie assembled an auspicious cast, which included Michel Piccoli, as an Andreotti-like president, Gian Maria Volonte as an Aldo Moro figure, a man of apparent calmness of mind, and Marcello Mastroianni as a reactionary priest (in the novel this character had been a painter but Petrie substituted a Jesuit in order, he claims, to 'place myself at the heart of the contradictions'). Petrie has remarked (*Jeune Cinéma* No 98) that although the priest is profoundly reactionary he is also the only one who 'ascends to a spiritual dimension less vulgar than the others'. He adds, however, 'I am convinced that anyone who believes profoundly in the moral principles of Catholicism can only be a reactionary because he conceives society in a vertical manner, founded on the practice of resignation and the God to Son axis.'

What emerges in *Todo Modo* is that, far from inducing in the voluntary exiles a new degree of tranquillity, isolation leads them towards their personal madness, a kind of 'morbid élan towards hell which is in us,' as Petrie describes it. In keeping with this vision, he presents the decors of the film as grotesque – garages like bunkers, a decimation of the natural environment, a sense in the protagonists of an invisible enemy, a halted situation in the outside world, a devolution. Petrie has said that when he made *Todo Modo* he was undergoing 'a feeling of satiety, of nausea, a drift towards pessimism'.

Todo Modo suffered a distribution eclipse by Warner Brothers, who initially held up the film's exhibition for political reasons and then issued it without promotion. Yet, in Italian political circles, the film stimulated controversy. Round tables were set up, made up not only of intellectuals but also priests, political journalists and Christian Democrat politicians, one of whom, Sarti, was particularly upset by Petrie's portrait of Moro, whom he defended as 'a man of great character and high intellectual stature'.

Petrie riposted that the portrait of Moro was gestural, showing a type, 'a modern paterfamilias, a hypocrite, a Catholic technocrat'. Sciascia himself, however, praised the film for being 'decidedly and justly anti-Christian Democrat. One could say that *Todo Modo* ... is the kind of trial of the ruling class which Pasolini wanted to make.' And which Pasolini did make, we might add, not only in *Salò* but also in *Theorem* and *Pigsty*.

Rosi's film of 1976, adapted by Rosi and Tonino Guerra from Sciascia's novel *Il Contesto*, with a cross-European cast including Fernando Rey, Max Von Sydow and Lino Ventura as Inspector Rogas, also grapples with the intensification of the social crisis. With the rash of shootings of judges during 1980 it becomes urgently topical again, as Rosi's films have a habit of doing.

We begin by watching the barely alive contemplating the dead. Varga, an elderly court Prosecutor, is shown walking through catacombs; his crumbling face contemplates the cracked sepulchres and coffins and, faintly, a dirge is heard, perhaps a sound in Varga's mind, perhaps a premonition. On the street, in sunlight, Varga stops to admire a hanging blossom. We have the impression that he is fortified by his tomb-mooching, and that this gesture towards a flower is not characteristic. A shot fires as Varga plucks the blossom. As he collapses clutching the tendril of honeysuckle a voice informs us, 'He talked to the dead to learn the secrets of the living.' We are now present at Varga's funeral, where a bunch of his legal cronies are paying their last respects, amid piles of garbage in the town square and crowds of demonstrators came to cheer the demise of one Fascist tyrant. Shortly afterwards, a second judge is murdered, on the *autostrada*, at which point Inspector Rogas is assigned to investigate the crimes.

Rogas, as a character, is a development in Rosi's pantheon of investigative heroes, not just because he is a policeman, but more importantly because he is not himself a man interested in power, like, say, Mattei, Luciano and Nottola. Instead, he represents Rosi's own intentions, that of the uncommitted investigator, in dramatized form, and serves to make coherent the ambiguities of the political fabric of the film.

In an interview in *Sight and Sound* (Summer 1975) Rosi told David Overbey that Rogas is the one character in the narrative initially exempt from corrupt or heavily compromised decision-making and power-mongering. But he adds, 'When a powerful man who believes blindly in justice begins to understand that he must doubt, the ground starts to slide from under his feet.'

The sliding beneath Rogas's feet is the element in Rosi's film which makes it more than another cop movie. The script carefully captures the man's initial steadiness of purpose. There is a touching scene in which Rogas interviews an old man with an allotment patch, whose land has recently been purchased by one of the murdered judges. Rogas strolls about admiring the peas; plucking one, he asks the old man what judge Sanza was like. The old man points to the distant city – 'like that, made of cement and smoke.'

Part of Rogas's poise is his willingness to listen, while remaining his own man. He becomes aware of a mystery man in the course of his inquiries, a pharmacist named Cres, one of several suspects who might have very good reason to hate the judiciary system; Cres, like several other suspects, has been to prison as a consequence of appearing in court before Varga. Rogas

Lino Ventura in *Illustrious Corpses*

is taken to Cres's house by a friend of his and discovers that Cres is a tango fanatic. Music pumps out from speakers in every room. But the house is now less a home than a mausoleum to a dead marriage, of which only the photos remain, those of Cres with the faces oddly excised from the prints. Rogas, convinced that Cres is his murderer, goes to see another judge, Rasto, to warn him that since he also served on the bench which indicted Cres, he may be the next to die. As he leaves, Rasto is shot in his bathroom by a rooftop gunman.

Rogas now tells a political journalist he knows that he believes the murderer is Cres, but when he puts his ideas to the police chief he is told that a Leftist conspiracy is the cause of the deaths. Rogas gives this notion short thrift – 'a campfire and a demo don't prove murder,' he retorts. He is taken off the case, and transferred to the political section, where the sort of authoritarian paranoia he himself has tried to resist runs rampant. Bugging and TV surveillance techniques are all installed here (a more sophisticated series of 'back rooms' than those suggested in Petrie's *Investigation of a Citizen ...*). Rogas is opposed to all this, but he does later make use of one surveillance technician, a friend, whom he asks to bug the police chief's phone.

Further machinations by the homicide officers lead Rogas into heated objections. The crunch comes when he is told to deny the evidence of his own eyes – he saw the chief of police, in company with the head of the armed forces, leaving an embassy – an event which the chief denies ever occurred. His confrontation with another judge, Riches (Max Von Sydow), is an additional disorientation. Riches is a fanatic who speaks of the absolute authority of the law in terms which chill Rogas. 'Today the only form of justice is decimation,' declares Riches, 'individuals do not exist.'

With Patto, the journalist, Rogas vists a Leftist party where young radicals quarrel. This is an important scene, showing the intense sectariansim of the extra-parliamentary Left, about which Elio Petrie has also commented disapprovingly. 'Their security is make-believe,' Patto tells Rogas. 'They're terrified.' He explains that the PCI now has to coexist with radical groups, just as it does with Christian Democrats. In this scene we observe the Historic Compromise in action.

Rogas himself is now under surveillance, as he discovers when he plays back the tape recorded from the police chief's telephone. He takes to sleeping in underground car parks, and in these rapid, nervy scenes we watch a man of poise go to pieces. Finally, during his meeting with a Communist Party official at the senate house, both men are gunned down by an anonymous assassin. The film ends, like so many of Rosi's films, with the crumpled bodies.

Rosi has altered Sciascia's novel in two significant ways: for Sciascia's mythical country he has substituted Italy; and for Sciascia's ending (in which Rogas himself shoots the CPI official) he has introduced the more plausible double assassination. This change of ending enabled Rosi to restore a political

The final assassinations in *Illustrious Corpses*

dimension to the finale, as he comments: ... 'the official version is that Rogas killed the PCI secretary and then himself – another manipulation of truth by the controlling powers.'

Rosi has, in fact, been criticized by the PCI for his political interpretation of events in Italy during this period. In an interview with Verina Glaessner (*Time Out* No 372), Rosi points out that the film was regarded as anti-communist; he declares, however, that 'I showed the tensions of this precise political moment ... how the centres of power were moving further from the citizen.' He also insists that, contrary to the anti-communist bias alleged by his detractors, he still sees the PCI as 'the only uncorrupt political party in power'.

Rosi's willingness to penetrate the texture of both Right and Left ideologies in *Illustrious Corpses*, along with his depiction of police interrogation and surveillance methods, make the film a valuable social document as well as a tautly orchestrated *film noir*. Not since *The Conformist* has a political thriller been quite so sparely constructed.

What the film demonstrates above all about Rosi's development is a new expansiveness of horizon. The denunciatory zeal of his earlier films, in particular *Hands Across the City* and *Just Another War*, is replaced by a more deliberate yet also more imaginatively concentrated method of presentation.

In Sciascia's novel, in the bleak little epilogue to the death of Rogas, two PCI members are discussing the frightening turn events have taken. 'The people will never know the truth,' concludes one of them.

For this line Rosi puts in another – 'The truth is not always revolutionary.'

Deportees of Mussolini in *Black Holiday*

The new sense of moderation implicit in this shift of stress is revealing.

When Carlo Levi wrote, in *Christ Stopped at Eboli*, of his year's internment in the villages of Grassano and Gagliano in 1935/6 he began by saying, 'Christ never came this far, nor did time, nor the individual soul, nor hope, nor the relation of cause and effect, nor reason nor history ... the seasons pass today over the toil of the peasants, just as they did three thousand years before Christ: no message, human or divine, has reached this stubborn poverty.' Levi was a man of culture from Turin. He was a writer and painter, but also a doctor of medicine. He was neither a Fascist nor a Marxist; he belonged instead to a dissenting liberal faction called The Justice Party. He is famous not only for his book, which has attained the status of a classic in Italy, but also for coining the phrase 'white telephone movies' to describe the tedious banality of many of the products of the Italian cinema during Mussolini's period in power. Dissenting liberals were almost as 'dangerous' to Mussolini as dissenting Marxists like Antonio Gramsci. Levi, too, found himself deported to the extreme south of Italy, although, unlike Gramsci, he was released after a year, following Italy's victory in Abyssinia.

Other Italian movies of recent years have touched upon this important (and hitherto somewhat overlooked) aspect of the Fascist era. Marco Leto's *La Villegiatura* (*Black Holiday*), of 1973, deals with a liberal dissenter *in confino.* Leto's political idol is Giolitti, a centre-left politician of that era, whom Leto has compared to a Moro in present-day Italian politics – a moderator who finally 'gets pensioned off', though of course Leto's remarks were made before the murder of Moro in 1979. Leto's film, however, differs markedly from *Christ Stopped At Eboli* in one significant respect, despite the strong congruity of theme and the fact that Leto incorporates details from real-life accounts of confinement, including a thorough reading of both Carlo Levi and Cesare Pavese's documentation of their experiences as detainees of Il Duce. Where Leto chooses to centre his drama round the polemic which develops between his Giolitti-influenced Professor and the communist and anarchist agitators who are detained on the island with him, Rosi has concentrated on the changes occurring *within* his hero.

The film begins with Levi's long train and bus journey to Gagliano (a village, in the film, which Rosi has 'synthesized' from two or three neighbouring village-locations, much as Ronald Blythe made a 'synthesis' of English hamlets to arrive at *Akenfield*). On the way he acquires a small, engaging dog, which somebody has abandoned at a railway station. He is now introduced to his new home, a clean, sparse shanty occupied by a taciturn peasant woman who will feed him, and suddenly finds himself a

Charles Simon as the priest in *Christ Stopped at Eboli*

marginal man in every sense. No culture, conversation, friends, and no politics. Gradually he meets the local people, among them the unctuous Fascist mayor, who is responsible for vetting Levi's letters and proscribing his reading list; a somewhat deranged but honest Catholic priest (marvellously played by François Simon); Giuglia (Irene Papas), a vital, erotic woman who has had seventeen illegitimate children, only one of whom appears still to be with her; and a number of lively American–Italian artisans and shop-keepers who have their memories of life in the New York ghettos known as 'Little Italies' to recount to the attentive stranger.

Levi paints, he walks (as far as he is allowed to – not outside the village outskirts) and he lies in an open (empty) grave in the small cemetery watching the blue sky and listening to the 'gravedigging' utterances of the verger. Of the other detainees (among them Communists, so he is told) he sees very little – only the plates of spaghetti left on a wall outside their hovels. Conditions and terms of detainment imposed on Communists were, generally, more severe.

It is this absence of intellectual contact about which Levi becomes acutely ambivalent. On the one hand, he finds it enormously irksome; on the other, he finds himself coming up against aspects of his personality he never knew existed. This crystallization of the self, which is at the core of Levi's account, Rosi and Tonino Guerra, in their adaptation, and Rosi and Gian Maria Volonte, in their screen realization, appear to have understood with considerable empathy. There are numerous moments of human vividness in the book which they have transposed gracefully to the film. One occurs after Levi has shifted and set up a studio room in Giuglia's house. He is painting at the easel and pauses to watch Giuglia's small son. His abstract gaze is so strange that the child becomes terrified, puzzling Levi, while Giuglia becomes intensely animated. Consistently, details from the text are fluently incorporated into the film.

The difficulty of adapting to the screen a work composed so much of interior monologue must have been considerable and some of the most trenchant of Levi's speculations about life in the south are under-stressed in the screen adaptation. One powerful motif in Levi's thinking, which Rosi/Guerra seem only partially to have drawn into the film, is his conviction that in the south there exists a matriarchial order, in contrast to his experience of life in northern cities, partly due to the widespread emigration of father-age males, both to cities in the north and to the USA. But more than this, there is, in the women who have remained, a sexual ferocity which is unnerving.

Rosi has focused the sexual vitality implied in Levi's comments upon Giuglia, who, in Irene Papas' playing, suggests some of this fascination which Levi describes. Like Levi, Giuglia is something of a marginal in this tight-knit community, and the locals regard her as a witch. Levi, however, perceives that what they see as sorcery amounts to no more than greater libidinal energy. He remains wary of Giuglia, knowing they could destroy one another. The film compacts this undertow of feeling between Levi and Giuglia into

Gian Maria Volonte in *Christ Stopped at Eboli*

one potent vignette: Giuglia is helping Levi bath himself. He is becoming aroused, she isn't. He mutters that perhaps he ought to leave. She becomes animated. He slaps her face. We are left to wonder whether they have been lovers, however briefly.

Levi, here, becomes a seasonal creature, rather than an appointment-orientated man of the city. He takes up doctoring, reluctantly at first, knowing there are two local 'quacks' who, although they ignore the medical needs of the peasants, are likely to resent any intervention on his part, particularly since he is charging no fees. He discovers, through contact with the working people, that although the landscape is harsh, the people are vital in their own way, able to sing, make music and play cards, and willing to let an honest man, however strange, participate in their pleasures. Levi learns, through them as well as through his internal changes, that 'In this ambiguous region ... everything is filtered through a sense of the sacred ... each thought ... each act of the imagination has the character of things first thought.' He implies that here there is nothing for a man to do but live. In a sense, Levi wanted, and Rosi now wants, to reveal this message to city-dwellers and those bound within any set of fixed assumptions, whether professional, cultural or political.

Reviewing the film in *Il Tempo*, Antonio Altamonte summed up the essential statement of the film as being concerned with the South not so much as a 'problem' (which has always been how the rest of Italy has seen all points below Rome) but as 'an initial condition'. Levi, too, makes a similar point when he says, 'What I call peasant civility is not limited to a region nor to men who work the land. We all have to discover in ourselves a little bit of Lucania.' Detained intellectuals appeared to undergo this paradoxical benefit; despite conditions of physical adversity, solitude cleared their minds and sharpened their perceptions. The classic illustration of this paradox is Gramsci's own situation; during his years of confinement in various prisons, he drafted his *Prison Notebooks*.

That Levi's work should be as much a celebration as a denunciation stems from the fact that most of those he knew in Eboli were not Fascists. One form of Roman madness was much the same as another, from their remote situation. Levi describes, for instance, the general absence of interest in the Abyssinian war – 'the peasants had no faith in a promised land which has first to be taken away from those to whom it belonged; instinct told them that this was wrong and could only bring them bad luck ... no one enlisted.' In the prevailing absence of imperialism he finds his own spirits lifting.

One of the key scenes in *Eboli* occurs when Luisa, Levi's sister from Turin, comes to visit him briefly. She brings with her the rationality of the north, the culture, and the attentiveness to social issues. Yet, Luisa's continuing belief in liberal notions of progress now strikes Levi as oddly redundant. Fond as he is of Luisa, he comments when she leaves, 'What she had told me of Italy seemed to belong to another period of time. I understood ... that her impression of the south was like a schoolboy's impression of Dante's Inferno.'

What is entirely novel in *Eboli* is that Rosi presents us with a hero who defines himself between two poles of feminine influence. There is Giuglia, with her stoical intuitive power, and Luisa, who represents the best elements of the world he has left behind – the curiosity about society, the wit and the compassion. Talking to David Overbey in 1976 about the absence of women in the majority of his films Rosi said, 'If women are not central to most of my films that is hardly my fault. I make films about power and ... women are rarely there ... when women become central to such things, as I hope they will, they will be in my films.' There is a slightly defensive tone here which is uncharacteristic of Rosi, suggesting that perhaps the absence of the feminine sensibility in his cinema is an unresolved aspect of his cinema. In *Eboli*, Rosi, following Levi, is able to contemplate this matriarchial community, and its impact on his hero, with an equanimity which, once, he would not perhaps have mustered.

Levi in the book arrives at a position which is essentially anarchist rather than social democrat when he says, 'the problem of the South will solve itself if we can create new political ideals ...' It is interesting to juxtapose Levi's notion about the irrelevance of the State to the real purposes of life with a comment from Giovanni Baldelli, derived from his book *Social Anarchism* – 'In order to appreciate the unethical element inherent in the State one has to compare it not with the rival powers it has supplanted but with what it has replaced in its alleged sphere of ethical actions. It has replaced habit and customs. Ethical people, even within the State, make efforts to revive old customs, to bolster up new ones that survive, or to think of new ones; they seem to realize that customs are the natural expression of the ethical will of society.' The world Levi depicts, and Rosi now restores to life thirty years later, is one in which an 'ethical people' (the peasants) express their ethical will in the simplicity and absence of self-aggrandizement of their customs. Levi, in Gagliano, realized the importance of the individual; no values can be derived except through the complexity of each person.

This introspective impetus to the book, partially coalesced via Levi's meditations about the unreality of all forms of institutionalized power, during a brief 'holiday' in Turin halfway through his period of confinement, Rosi and Guerra are hard put to convey in the film form. In a sense, though, Rosi's dilemma was eased by virtue of a curious production arrangement made available to him by RAI TV, who financed the project. In addition to preparing a cinema version, running some two and a half hours, Rosi also prepared a longer version for television transmission (in four parts). This version, which contains the book's more didactic elements – Levi's

Gian Maria Volonte in *Christ Stopped at Eboli*

conversations in the Turin library with Luisa and friends – runs some four hours.

Finally, it should be said that there is a lyricism about Rosi's *mise en scène* which should not be entirely surprising to those who are familiar with *C'Era Una Volta* and, to some extent, *Moment of Truth*. *Eboli* makes much use of blue skies; there is also Bertoluccian energy to the camera movements – the rapidly panning descent of the hills, for example, after the radio announcement that Abyssinia has capitulated, showing unconcerned peasants working in the fields and finally ending in the waste lands.

When I talked to Rosi in Rome in June 1979 he observed that he sees *Christ Stopped At Eboli* as a tribute to a way of life which has largely vanished in the South of today, in which educational facilities have been made available, sanitation has at least partially eradicated the once endemic problem of malaria, but in which unemployment still remains a major problem. He pointed out that, following the screening of the film in Italy, a round table conference was convened in Mattera, at which politicians of the PCI, Christian Democrat and Socialist parties, along with peasants and trade unionists, were able to debate the issues raised in the film concerning the isolation, economic and cultural, of the South. He also commented that the mass exodus of young workers into cities like Turin and Milan, as well as across the border into Germany, Switzerland and Belgium, has had a devastating effect on the North, too. Turin, for example, Rosi describes as 'the largest Southern city in Italy'.

Above all, though, he insists that the essential theme of Levi's book is contained in his injunction to 'do one's work lovingly, and be content'. Rosi himself has always 'done his work lovingly' but the very particular flavour of *pacatezza* in his own work at this time suggests a measure of equilibrium which should bode well for his future. To approach Rosi's present theme of *pacatezza* (a word which he insists carries no precise English equivalent but which has the approximate sense of 'tranquility') we now broaden Rosi's perspective on the *contadini* to include other notable examples of Italian cinema's renewed interest in agrarian themes, in the following chapter.

6. Peasant Perspectives: Ermanno Olmi and The Taviani Brothers

Ermanno Olmi was born in Bergamo in 1931 and, after attending an acting course at the Milan drama school, began his career in cinema working on short documentaries at Edison. The film which bridged the jump from his thirty-seven documentaries for Edison to his feature film career was *Il Tempo Si E' Fermato* (*Time Stopped*), which he wrote and directed in 1959. *Time Stopped* is a study of the relationship between two men, one old and one young, who meet on a snowy mountain. The film has its area of social significance, for one man is from the South, the other from the North, and their relationship reveals the cultural clash which has been so evident in post-war Italian cinema. But it is essentially, like all Olmi's films, a poetic celebration of the moment of contact, the interruption of the normal course of events, the recognition of a common humanity which reasserts itself once man has the opportunity to move away from the divisive tensions of city living and the industrial process which ordinarily oppresses him.

In 1961, Olmi consolidated the independence he had gained with *Time Stopped* by forming his own production company, in partnership with an intermittent collaborator of his, Tullio Kezich, the purpose of which was to provide an umbrella not only for his own future productions but also for other film-makers seeking a start. The company was called '22nd December' and was responsible for some half-dozen movies before it folded, among them Lina Wertmuller's debut feature, *I Basilisci* (*The Lizards*), and Damiano's *La Rimpatriata*, both films dealing with the difficulties of life in the rural South. In the same year he made his second feature, *Il Posto* (*The Job*), a film which, in its control and delicacy, is an ample testament to the value of Olmi's long apprenticeship making short documentaries for Edison. The film shows the sense of wonder and bafflement of a youth from the suburbs who arrives in Milan for a job interview, and subsequently begins commuting, having been hired as an office boy in the large, impersonal organization.

His sense of wonder is revealed most acutely in the walkabout scenes around the main Milan thoroughfares, which he makes in the lunch-hours, in company with a girl who has also just started work in the same organization. They marvel at the smell and taste of strong coffee, gloat over the expensive clothes in the stores, and run through the traffic and the parks, holding hands to protect one another, arousing the ire of an officious park-keeper. The youth's bafflement emerges in the many scenes set within the organization. He is bewildered by the quarrelsome clerks, not understanding that they are frantic because there is not enough to do, rather than too much. His only friend is the porter, whose phlegmatic manner suggests he has seen it all. Together they watch the company employees from the porter's tiny cubicle at the end of the corridor. The girl he hardly ever sees for she has moved to another department. During one fleeting tryst in a corridor they tentatively arrange to meet at the company's New Year dance. But the girl is not present; the youth sits alone, eventually being taken over by a maternal older woman who persuades him to dance with her.

The absence of posturing, the clear observation and the dry humour of *Il Posto* made it a favourite with English audiences, reviewers and young film-makers in the early 1960s, as is revealed by Peter John Dyer's appraisal in *Sight and Sound* (Winter 1961/2): 'The film remains a gem of understatement, with the warmth, freshness and human oddity of the best social comedy ... few films have seemed less bitter about futility. Olmi is one of the rare talents who can apparently transfigure poverty of scope and ambition without sentimentalizing it.'

What Dyer's assessment does not quite suggest is the degree of formal originality which pertains in the movie. Having set up an expectation of romantic interest in the early sequences of boy meets girl, Olmi then adroitly draws the narrative along quite another path. The girl hardly ever meets the boy, and the New Year's party also confounds expectations, for again they fail to meet up. Instead, Olmi shows the destructive effects of office routine. During one clerical scene, he suddenly cuts to a mute sequence showing one clerk at home – where the working pattern of isolation is repeated. This abrupt side-slip is vindicated, in narrative terms, when later the clerk dies, and a scramble begins among his colleagues to grab his well-situated desk, at the front of the office. There is something of Melville's *Bartleby* or Kafka's *The Castle* in these depictions of clerical hypermania; perhaps the only other European director who has captured office tantrums so tellingly is Claud Goretta, in *The Invitation*.

With his third feature, *I Fidanzati* (*The Engaged*), made in 1963, Olmi reversed the poles of the ever polemical North–South debate by showing the shift to the South of a Northern petro-chemical worker who goes to work for a firm in the South, leaving his fiancée behind in a Northern city. The film is a study of the isolation of the young engineer, in his working and domestic life in the small town. Intercut with these scenes are love letters exchanged between the engaged couple, in which they express their frustration at being crudely separated. Olmi has said that he wanted to show

how, in this Southern town, everything has slowed down while 'the only ones who continue are the great petro-chemical firms which do not have, so to speak, the privilege of pause ... this "grace", this pause, for me is basic.'

The neo-realist simplicity of Olmi's work, so evident in this comment, echoes quite directly Roberto Rossellini's observation that, for him, the key to artistic authenticity lay in the waiting, the patience, the avoidance of frenzy.

In the mid-Sixties, Olmi turned to television, completing several documentaries for RAI on religious themes. These included *St Anthony* in 1964, *Dopo Secoli* (about the pilgrimage of St Paul) in 1965, and *Storie di Giovanni* (a six part documentary series about Pope John) in 1966/7. These were followed by further documentaries, but of a social rather than religious theme – *Cento Anni di Galleria* and *Queste Estate: Ritorno al Paese* in 1967, and *Chi Legge in Italia?* in 1968. That same year, however, Olmi returned to features with *Un Certo Giorno (One Fine Day)*.

This film shows the crisis which occurs in the life of an advertising executive named Bruno, whose machinations at work form the early part of the drama. Bruno begins an affair with a sultry market researcher, but is suddenly brought back to earth when he is involved in a mysterious road accident: he knocks down and fatally injures a roadmender, and finds himself standing trial for manslaughter. After his acquittal he returns to his wife and family, much chastened. Olmi is concerned to show the intricate connection in the man's mind between his adultery and the 'punishment' he sees himself as receiving. Again, the 'moment of pause', in this case a long night Bruno spends alone while awaiting a police interrogation, is crucial to the story's meaning. During this hiatus, a shift of values takes place in Bruno.

Bruno's world is briefly and somewhat schematically contrasted with that of a young painter, the boyfriend of the market researcher, who lives humbly in a garret and offers the girl less sensual velocity than Bruno but far more serenity. A similar juxtaposition is set up in Olmi's later *The Circumstance*, as we shall see shortly.

Olmi himself sees *One Fine Day* as being crucial to his artistic development. As he commented in *Positif* (No 185), 'I had arrived at a period of maturity. I was seeing clearly on the theme of responsibility. I abandoned the script, I found myself with a story which presented itself as a kind of primitive spectacle, and thus rediscovered great liberty.' The growing impact of Olmi's increasing emotional range becomes even more apparent in his choice of subject and treatment in all three films which follow.

I Recuperanti (The Scavengers), of 1969, was adapted by Olmi and Tullio Kezich from a book by Mario Stern, and set at the end of the Second World War. The story concerns a former POW named Gianni, who returns to his village in Northern Italy to find that his father has remarried, his fiancée Elsa has altered, and little work is available. His efforts to start a cooperative are vetoed by the local council and he becomes involved instead with an elderly anarchic individual named Du, who works in the hills recovering metals for a railway company – risky work, since bomb fragments are common. Gianni learns how to take soundings for lead, iron, bronze and copper; he helps defuse a bomb (called 'killing the pig') and when he eventually returns to the village to build himself a house Du accuses him of abandoning him. Gianni promises to go back some time.

The Scavengers returns to the theme of *Time Stopped*, stressing patterns of work which occur outside the norms of industrial society. Marginal activity is more likely to restore the sense of space and the moment of pause, Olmi suggests. He has described the film as 'a war under a man's skin, but also under the earth's skin – dig and you find bombs, and bodies. War is, at bottom, the moment of explosion which doesn't detonate.'

During the Summer, of 1971, which Olmi co-scripted with Fortunato Pasqualino, and again made for RAI TV, studies an odd, other-worldly relationship between a middle-aged cartographer and a girl he meets in Milan, quite by accident, when they jostle on a traffic crossing and she screams at him and runs off, arousing his curiosity. They meet again, and become friendly, sensing that they are both embattled, he with his employers and she with stray boyfriends whom she always ends up leaving after violent quarrels. For the Professor, absence of colour sensitivity in the world around him is a great tribulation. Colour is everything, he tells the girl during an outing to a floral exhibition in a formal garden. He even resigns his job when his employers insist that he colour Tuscany (on a map he is preparing) in blue not yellow.

For the girl, life's difficulties are compounded by irate boyfriends. 'First they pretend they don't care and then they get mad with jealousy,' she confides in the Professor. 'My feelings just come.' She wonders if she is a slut; the Professor gently reassures her that to him she is a princess – 'there will always be princesses, until every man discovers his own nobility.' To aid mankind in this process he selects passers-by on the street and sends them family crests he designs himself with notes congratulating them on their hitherto unknown noble lineage. He is eventually arrested for fraud and, at his trial, the girl speaks for him as a witness. She tells the attorney that she is a princess. 'In that case,' he retorts, 'I'm His Majesty.' She replies coolly, 'No, you are not even a gentleman.' The Professor's respect has enabled her to find the poise she requires to defend herself in a man's world.

One critic, Eithne Bourget, in an article entitled The Art of the Heraldic, has described the colour symbolism of *During the Summer* as expressive of 'an eccentric humanism common to the works of Christian myth'. In fact, this film mainly reveals Olmi's deep instinct and reverence for the harmony of the natural order.

The film is distinguished particularly by its sexual reticence and its extraordinary editing fluency, which lends great conciseness to a somewhat fey storyline. The Professor and the girl have an entirely platonic relationship, almost quaintly chivalrous on his side, and maidenishly demure on hers. They

Rod Steiger in *There Came a Man*

enact with one another a model relationship of courtly love they could find nowhere else in Milan, and it works for them. In the final shot we see the Professor behind bars calling goodbye to the girl as she leaves the courtroom, now able to face the world.

The most complex and assured of these RAI TV movies is *The Circumstance*, of 1974, which returns to the main theme of *One Fine Day* – the executive involved in a road accident, and the potential change of values induced by this act of God. The story concerns a hard-working actuary, a woman with a husband made abject by his fears of redundancy (he is an executive in a corporation marketing business games) and two grown-up children. She witnesses a terrible accident while driving to her summer mansion. A young motor-cyclist crashes into an overturned caravan and catches fire when his petrol spills. She rushes him to hospital badly burned. Much moved by his terrible plight, she returns at intervals to see him as he lies heavily bandaged in bed, hardly able to speak, awaiting the arrival of his parents.

The woman's secret involvement with the burned youth is counterpointed with her family relations. Her teenaged daughter is falling in love with a boy she has met at the private swimming-pool in the mansion grounds. Her son and his young wife are expecting a baby. Her husband needs continual reassurance about job security which she, as an entrepreneurial fixer, is supposed to provide for him. All these elements are beautifully meshed, and

193

given fairly even prominence. This was not the case in *One Fine Day*, where Bruno's family relationships were somewhat underpowered.

There is one scene especially which epitomizes Olmi's ideas, and prefigures the themes he will explore more expansively in his later *L'Albero degli Zoccoli*. In a violent thunder storm the young wife in her outlying villa goes into labour. Her husband cannot get the car out of the mud and has to knock up neighbours. The baby is delivered safely at home and, although the doctor is sanguine enough about his home delivery, telling the young father, 'We have lost our faith in life,' the new grandmother, when she arrives, is far less happy. She wants her son and daughter-in-law to move back into the mansion, largely for the sake of appearances – she doesn't like the prospect of her executive ladyfriends calling at this simple abode with their baby gifts. Thus Olmi shows that the woman's attachment to the boy in hospital is basically a sentimental disfigurement; she has not changed very much. Social appearances still define her. And meanwhile her younger child, Sylvia, is becoming estranged, unable to discuss her developing sexual feelings. Olmi shows all these characters locked in a vacuum. It is an affluent milieu, but sterile, and only the young couple with their new baby have broken free to achieve contact with the life force.

In *The Circumstance* Olmi says goodbye to his consistent preoccupation with modern industrial working conditions, using a beautifully controlled irony: a number of brief scenes show the anxious executives taking part in a business game. The object of this is to increase their competitive cutting edge. But the more fun they should be having, the more morose they become. Offices are mysteriously vacated. Desks are cleared overnight. Redundancy hangs like a cloud over each one of them. In this potent crystallization of failing industrialism Olmi prepares the ground for his developing curiosity about alternatives to corporate industrial life. In 1978 he made *L'Albero degli Zoccoli* (*The Tree of the Wooden Clogs*).

Olmi once remarked, concerning *I Fidanzati*, 'When one tries to avoid being bound to real time, one sees how temporal logic is derived from the present moment alone, but is continually compromised by past and future.' By stepping back in time, to Bergamo at the turn of the century, he shows how the present era has been 'compromised' by the past, and how this may affect our future. Using local farmers as his characters, he as usual shot, scripted and edited. His film shows peasants who, by their nature, do not participate in society and culture, but provide instead the rich compost in which others nurture ambition, belief in progress and material goods.

The peasants are aware that they are excluded from these processes but there is no surplus energy to come to conscious terms with the fact of exclusion. The landowner for whom they work takes two thirds of all they harvest.

Submission, then, is a prime characteristic of the community depicted in *The Tree of the Wooden Clogs*. When the farm bailiff evicts Batisti and his wife and three children none of his neighbours protest against this injustice, though to a man they feel it. Batisti's crime is that he has felled an alder by the river to fashion a clog for his son. This absence of solidarity which Olmi shows is the antithesis of the agrarian solidarity which Bertolucci depicts in *1900*. There is little evidence of militancy in Olmi's film, only the socialist orator who hectors the crowd during the carnival sequence, and the pall of smoke on the Milanese skyline as Maddalena and her new husband approach on a river barge. The file of chained agitators who are marched past them in the streets as they arrive makes the point: dissidence is brutally repressed. Olmi has no consoling red flag to offer.

Yet there is solidarity of a different sort in Olmi's peasant community. It stems from a common sense of affliction but also a common sense of occupation. From the apparent passivity arises a spontaneous conviviality, in the evenings around the hearth telling stories, and in the summer working in the fields together, sowing, harvesting or wood-gathering. The peasants ritualize their hatred towards the owner class, just as Carlo Levi noted the Catanian peasants doing during his exile in their Thirties. A story which Batisti tells, for example, is full of lively malice towards the owner class: a man breaks into a rich woman's tomb and steals her jewellery. When he cannot prise the ring off her fingers he cuts off her hand. Later, driving at night, he gives a lift to a sombre woman. His horse bolts to the cemetery, where she wishes to alight. As he gets down saying 'give me your hand' she replies, 'You already have it.'

The story provokes peals of laughter from the gathering. Implied in this laughter is a form of fatalism. Even if you try to put one over on the moneyed classes, they always get their own back – even, perhaps, beyond the grave. Batisti's story is prophetic, for he does subsequently steal the tree, and is evicted. Guile in other contexts, too, is unrewarded. When a peasant finds a gold coin at the carnival he runs home to secrete it in his horse's hoof. When he goes to retrieve the coin later, and finds it gone, he attacks his horse. The animal is so amazed that he attacks back, and the scene ends in farce, with the terrified peasant cowering in his hovel.

It is interesting to note how Olmi realizes these scenes which show peasant connivance. Batisti, stealing the tree, and the neighbour concealing his coin, use mud to hide the sawn stump and the wink of gold within the horseshoe. (Soil becomes a means of beating the bosses.) Both acts of guile are presented without music. Direct sound alone intervenes – the scrape of tools, the heavy breathing. Elsewhere, direct sound, particularly church bells, regularizes the assorted daily activities of the peasants at work. Moments of heightened activity, on the other hand, are underpinned by Bach organ or cello passages – the boy setting off to school on the momentous first day carrying that symbol of status which might make his future so different from his parents – the satchel, or the washerwoman running to the church to fetch water with which to bless her sick cow, muttering a continuous incantation. The most compelling tension in the film between the routine and the transcendental occurs, however, in the observation of the young couple.

Peasants at work in *The Tree of the Wooden Clogs*

The boy and the girl, in their tongue-tied encounters in the fields and the barn, their marriage, and the honeymoon journey, are crucial to Olmi's intentions, as indeed were the relationships of the young couples in *The Job* and *The Engaged.*

Within the farm community, the young couple's lassitude differentiates them from the voluble world of their elders. The older men, Batisti and his peers, and Anselmo, the grandfather of the washerwoman's children, are highly articulate, witness their storytelling flair and the ready wit of the men on the morning of Maddalena's marriage. Why this contrast? Are these young people so silent because they are filled with wonder? Novelty is part of their life, as it was for the boy and girl starting work in a large company in *The Job.* They get married at a moment in history when certain great changes are about to erupt: in the scenes in Milan, Olmi shows us the first stirrings of working class agitation. These two bridge the divide between the old forms of oppression – exemplified by Batisti's eviction and the washerwoman's destitution – and the future: technology, mass affluence, city alienation.

The scene which crystallizes Olmi's intentions is the wedding journey, by barge, to Milan. It is a strangely festive journey, during which the travellers picnic, bourgeois beside peasant, and all share the common vista, which includes the pall of smoke upon Milan's skyline. There is an exuberance, visually and acoustically, about this scene (the shift from organ to cello music signals Olmi's intentions clearly) which exists nowhere else in the narrative. The future encroaches, and new customs are hinted at, in which people of different social categories may sit together and share certain celebratory yet also threatening experiences. In this scene, so to speak, Olmi allows the twentieth century to arrive.

The delineation of the scenes at the convent is curious. What, for example, does Olmi intend in his facsimile of The Last Supper composition, in which twelve sit at high table in the convent dining room, ten nuns and the newly weds? Is this a First Supper, in which the bride and groom replace Judas? A supper before the fall from grace, then? Two scenes which follow make Olmi's purposes a little plainer. The couple are taken through a dormitory of sleeping foundlings. 'They're really little angels,' says the Mother Superior, 'the world is a better place in here with them.' This reverence for the innocence of children echoes the comment of Batisti's wife when she reproves Batisti for his fears on the birth of their third child, 'Don't worry. God gives every new born baby a small bundle of his own.'

The bride and groom are left alone in the 'bridal suite', a single room with two dormitory beds pushed together, the join concealed with a garland. Olmi discreetly elides the carnal action, resuming on them the next morning. She is braiding her hair, as demure as before, apparently unmarked by passion. He is gazing out of the window. 'So many roofs,' he mutters. Maddalena's aunt, the Mother Superior, comes to offer them a foundling. If they take the child, she explains, the convent is empowered to offer certain small payments, until the child is fifteen. The couple hesitate, perhaps unsure about taking on such large responsibility so soon. The nun clinches the transaction by putting the toddler on the bed and pushing him gently towards Maddalena. One step, and the girl opens her arms. The decision is made for them.

The sexual reticence of Olmi's portrayal of the young couple is characteristic. Gavin Millar, reviewing the film in *The Listener* (3rd May 1979), referred to Olmi's 'daintiness' in this context. What the daintiness masks, however, is a powerful urge to be conjoined, which is evident in the courtship scenes, when the young man follows the girl home from her place of work, the spinning shed, and emerges panting in front of her, saying, 'I would like to give you a kiss.' She says, impassively, 'There will be time for that later.' This is a strong erotic moment. The act, left unspoken, becomes vibrantly understood between them, echoing similar moments of yearning between young couples in the early films. It is the minimal expressive range of Olmi's amateur actors, often so effective, which here tends to reduce emotional delicacy into mere 'daintiness'. This is true of the morning after the honeymoon scene; in the girl's face, nothing has altered. Olmi's elisions amount to a gesture against the phallocentricity of modern cinema. It is here that the charges of passivity which have occasionally been levelled at Olmi become most real.

One element in *The Tree of the Wooden Clogs* which is unique to Olmi's study of turn of the century peasantry is the role versatility of his peasants.

The two main male characters, Batisti and Anselmo, are shown as gentle, humorous men, with the expressiveness of natural artists. Anselmo raises most fully the role flexibility of the peasant. He serves as an all-purpose supportive figure in the washerwoman's household. He is not only the entertainer, but also the nanny and market gardener. His sense of guile is devoted wholly to practicalities. He plants his tomato seeds a week early in order to beat the rest of the harvest at market. His granddaughter aids and abets him. Anselmo, at one with himself, able to do a bit of everything, is a man who can cope, whatever is thrown at him.

Equally, Batisti's wife is shown as fully independent as a disciple *and* decision-maker within the family. Batisti is dubious about her decision to forgo the midwife (to save the fee) but is overruled. His wife feels convinced that her neighbours will see her safely through the birth, as indeed they do. As the women leave, after the birth, one says to Batisti, 'We have to help

The Tree of the Wooden Clogs

each other.' This is a sentiment from the heart of Olmi's humanism, which is also of course the humanism of the neo-realist cinema out of which Olmi's films have developed. Compare it, for example, to Nino's remark, at the end of Visconti's *La Terra Trema*, when he realizes the fishing coop has failed and says, 'We want to let everybody work so they can earn their own bread! We want to help everybody!'

The only scenes in which a patriarchal position is established are those showing the relationship between the washerwoman and her son, in which she defers to his decision-making ability once he starts work at the spinning shed, thus becoming the main breadwinner. This shift, from callow youth to head of the family, is in fact a manoeuvre on the part of his mother to increase his still frail confidence, and it is part of Olmi's subtlety that he denotes this relationship in terms of the mother's willing compliance.

Certain English reviewers have seen *The Tree of the Wooden Clogs* as a vision of an Eden, a place where The Fall has not yet occurred, yet Olmi's religious faith is arguably less devout. There is a contrast set up between the simplicity of the peasants' faith and the fracture imposed on their coherence by material stress. Olmi's message is clear enough, none the less so for being unemphatically delivered: profit consciousness is repressed and divided consciousness. Anselmo says, for instance, 'There used to be lots of little hungers. Now there is one big one.'

This remark seems prescient, given what we know of contemporary peasant communities such as those reported by John Berger in his essays, and his book *Pig Earth*, and certain Swiss documentary film-makers like Fredi M. Murer and Remo Legnazzi, who in their recent documentaries *We Mountain People* and *Chronicle of Prugiasco* show how threatened present-day rural communities are in the Alpine regions of Switzerland and Italy. These documentaries on Swiss mountain peasants reveal, like Olmi's feature film, the extent to which peasants depend on cooperation to survive, both inter- and intra-familial. Thus, the women work alongside the men in the fields, but stop earlier to prepare supper; while all set about certain big jobs, like haymaking and log-gathering.

It is probably no accident that Olmi's film should be so well received (taking the Critics' Prize at Cannes in 1978 and being almost universally accoladed by reviewers in France, Italy and the UK), for it comes at a time when general belief in progress, along the existing model of ever increasing technological stimulation, is waning. Implied in Olmi's film is a statement which has been made explicit by John Berger in his essay *A Class of Survivors* (*New Society* 22nd December 1977): 'Closely connected with the peasant's recognition of scarcity is his recognition of man's relative ignorance ... he never supposes that the advance of knowledge reduces the extent of the unknown. To believe that seems to him naïve.' This statement is significant in the way that it connects with Olmi's film. Olmi, like Berger, does not see the peasant's absence of technological stimulation as an impediment which bars him from 'the fruits of progress'. Rather, he sees the peasant's

The Tree of the Wooden Clogs

ability to make and mend as a vital factor in his resilience. He does not senti-
mentalize the farmworkers of the Bergamo community – we see them filching
one another's wood as well as the farmer's – but what he accomplishes most
cunningly is a 'period' film which is in essence a prophetic statement about
the post-industrial era now dawning.

The significance of this shift in attitude towards the peasant can be illustrated by recalling a film about peasant values made twenty years ago,
Bolognini's *La Viaccia*. In this the tenant farmer is shown as avaricious,
brutal and narrow-minded, hence the conflict which develops between him
and his fractious, Florence-bound son, Amerigo. In the ensuing two decades,
attitudes towards the practitioners of the simple, self-sufficient way of life
have altered. Far from being the greedy hick, the peasant is now focused
as critically relevant to current social conditions. His resistance to the hysteria
of technological and bureaucratic over-stimulation is seen as an advantage.
John Berger sees a cruel irony in the fact that we are coming to accept the

relevance of self-sufficiency values at just that point in history when they are about to vanish. Olmi, too, perhaps accedes to this point of view. The tone of *The Tree of the Wooden Clogs* is, after all, elegiac. It is worth noting, however, that the most useful film yet made about the beginnings of *new* peasant values, in which a small group of alienated professional people come together in a Swiss small-holding, was co-scripted by John Berger. This film, *Jonah, Who Will Be 25 in the Year 2000*, shows the inception of post-industrial peasantry, in which the common sense of survival in adversity is based, not on scarcity values, as the peasant domains of Carlo Levi's Catania in the 1930s or Olmi's Bergamo in the 1900s are, but on *surplus value*. Too much of everything drives the *Jonah* 'clan' into retreat – too many goods, too many people, too much harassment, too many regulations, and above all too much ugliness in cities.

What is markedly lacking in contemporary Italian cinema is any clear sense among the film community of the growing importance of this social and psychological theme. Olmi, with his *Tree of the Wooden Clogs*, has come closest to grasping the nettle. Pasolini, had he lived, with his uncanny prescience for drawing out the hidden pains and pressures of his country, might also have begun to penetrate this vital area.

At this juncture it is pertinent to introduce the cinema of the Taviani Brothers, Paolo and Vittorio, for they, in their eight feature films made since 1962, have shown consistent interest in themes of utopian import, and have frequently signalled their preoccupation with the role of peasants in this scheme. Unlike Olmi, however, they present this preoccupation in terms of the rhetorical political gesture rather than the closely observed realism which is the hallmark of Olmi's cinema.

The Tavianis were born and brought up in rural Tuscany. Like that other Tuscan *cinéaste*, Mauro Bolognini, they show in their films the mixture of visceral passion and the love of operatic gesture in their scenarios which they attribute to their relish for Neapolitan rather than Roman culture. At the same time, they are intellectually attracted to the agrarian anarchism of Russian political theorists, notably Tolstoy and Kropotkin. Their films are thus a curious *mélange* of red flag waving and cerebral distancing devices in the manner of North European theatre exponents such as Meyerhold and Brecht. They have always worked together, their views, as expressed in numerous interviews, most notably with Jean Gili (in *Le Cinéma Italien*), are interchangeable, and they regard their working relationship as far more than a collaboration (of the sort they had with Valentino Orsini, who co-directed their first two feature films with them) and more in the region of a 'blood brotherhood' – what they term a 'way of breathing'.

Their eight films to date are: *Un Uomo da Bruciare* (*A Man for Burning*) 1962, *I Fuorilegge del Matrimonio* (*Marital Miscreants*) 1964, *Sovversivi* (*Subversives*) 1967, *Sotto il Segno dello Scorpione* (*Under the Sign of Scorpio*) 1968/9, *San Michele Aveva un Gallo* (*St Michael Had a Rooster*) 1971, *Allonsanfan* 1973/4, *Padre Padrone* (*Father Boss*) 1977, and *Il Prato* (*The*

Meadow) 1979. With the exception of *Allonsanfan* and *Padre Padrone*, none of these films has as yet been made commercially available in the UK. Before 1962, the Tavianis also made a number of documentaries for RAI, among which was *L'Italia non e un Paese Povero* (*Italy is not a Poor Country*) in collaboration with Joris Ivens. Their cinema, like Olmi's, has always been a highly personal medium of expression, with the exception of *Marital Miscreants* (an episode film which presented different cameos of the absurdities of the then Italian divorce laws – and which the Tavianis now exclude from retrospective seasons of their movies), and has depended on the good faith of a persuasive producer, Giuliani G. De Negri, finance from RAI's second channel, and the unpaid help of friends, students and workers to achieve production.

The Tavianis' output seems divided evenly between films with a contemporary setting and films with an historical basis. Yet even those films nominally attached to present-day realities, *Un Uomo da Bruciare*, *Sovversivi*, *Padre Padrone* and *Il Prato*, have tended to treat the present era through a filtered perspective. The remaining films have dealt with revolutionary leaders or revolutionary groups during the nineteenth century and the earlier part of this century. The Tavianis describe their cinema as an onrunning project of research into utopia, asserting that they deliberately oscillate within each film from a position of intellectual detachment to dramatic situations demanding intense emotional involvement, thereby effecting a push-pull relationship with their audiences as well as themselves.

Un Uomo da Bruciare, the Tavianis' debut feature of 1962, starring Gian Maria Volonte (also in his cinematic debut), was based on the career of a trade union leader, Salvatore Carnevale, who was murdered by the Mafia some years earlier. The style of the film is thus very different from other studies of Sicilian militants of that time, Rosi's *Salvatore Giuliano* or De Seta's *The Bandits of Orgosolo*, being strongly theatrical in approach. Rosi's *Salvatore Giuliano* is conspicuous by his absence in Rosi's inquiry. The Tavianis' hero, however, is present in almost every scene, haranguing the workers in verse from his rostrum, envisaging his own death in a series of premonitory flashes.

What the Tavianis have to say about their first film indicates the basic format of their enthusiasm, both artistically and politically. 'For Salvatore,' they have said, 'union activity has the same worth that creative work has for the artist ... discovery: a hard task, contradictory, even dangerous, but one which obliges a man to relate to affairs, and to other men ... to become not a spectator or a victim but a protagonist.' The Tavianis, then, exhibit a relish for the world of political power in the way that Rosi also does, but where Rosi wishes to dissect these processes, the Tavianis wish to orchestrate them, so to speak, to show both the ardour of utopian hope and the ferocity of that hope when frustrated. There are set-piece scenes in their political films of the late Sixties which demonstrate this intentionally operatic style of the brothers' political vision, but before we discuss these scenes, let us consider briefly their 'rogue' movie, *Marital Miscreants*, which is somewhat less irrelevant to the Tavianis' cinema than has been suggested by critics.

This film has been described by Tony Mitchell (*Sight and Sound*, Summer 1979) as in 'a vein of light comedy rather remote from the serious political inquiry contained in the Tavianis' subsequent work'. This verdict seems to me to accept the film along the lines of the Tavianis' own retrospective dismissal. In fact, it is sometimes the case that the minor, apparently maverick work in a director's *oeuvre* proves the most revealing as to his unconscious intentions and enthusiasm. In the cinema of Rosi, for instance, *C'Era Una Volta* occupies this role, revealing Rosi's interest in Neapolitan music-hall and fairy tale.

Marital Miscreants, then, the one film in the Tavianis' cycle of thematically linked political films which stands apart (being based on a book by an Italian politician Luigi Sandone's abortive efforts to reform the Italian divorce laws), reveals their capacity to handle comic and fantastic elements which subsequently manifest in the films which the Tavianis see as seriously political. We see a GI bride who has been abandoned by her American husband, failing to gain her divorce in order to remarry; we see a mental hospital inmate mistake her husband for an authoritarian psychiatrist; and we see a man and a woman talking, against a plain grey background, about the impossibility of any love-bond, given the rigidity of the law, without ever meeting. In short, the film reveals a vein of absurdist humour.

Sovversivi, of 1967, derives, like *Un Uomo da Bruciare*, from documentary material which was shot before the feature film became developed. In this case, the documentary basis was the funeral of Palmiro Togliatti. The Tavianis participated in a collective film, footage from which they spliced into the feature, which concerns five individuals 'who seek to burst out of apparent tranquillity', in this respect echoing the Tavianis' own sense at that time of 'not wanting to suffocate', as they expressed it to Jean Gili. 'We felt a need, which was above all physiological, for shattering this pseudo-equilibrium. How? There existed no mass movement.'

What form does this urge for change take? For the dissatisfied wife of a Communist official, it is sexual; she becomes a lesbian. For her husband, change consists of coming to terms with his wife's assertion of independence. For another Communist, a young photographer, change consists of breaking his family ties and developing his career ambitions, while proclaiming the imminence of the Communist revolution in Italy. For another young man, change means leaving Italy to take up the political struggle in his native Venezuela. And for a film-maker called Lodovico, who is having trouble completing his documentary on Leonardo da Vinci, change means reorientating his film. He abandons the attempt to show the solitary genius of the artist for a more 'liberated' rapport with the people, declaring that 'Art is not enough – it repels you.' The film ends on a shot of Togliatti's coffin being lowered into the grave, suggesting a moment of truth, the death of the father.

The theme of suffocation takes on a more awesome physical dimension in the Tavianis' next film, *Sotto il Segno dello Scorpione*, depicting the evacuation from a volcanic island of a party of young revolutionary refugees. They land on another island, where a hierarchy exists: the community comprises farmers who coexist under a former revolutionary, Renno, now self-styled king of the community. This island, too, is volcanic, and the 'Scorpionists' try to persuade the residents to depart with them for the mainland. Their failure to persuade the residents to fall into line with their hopes for solidarity leads the Scorpionists into a frenzy of violence, in which the women are raped and the men massacred before they depart for the mainland.

Like Comolli's Franco–Italian co-production, *La Cecilia*, *Scorpio* raises the crucial question all those with an utopian predisposition need to ask themselves: is non-hierarchical, non-authoritarian communality feasible in practice? Like Comolli, the Tavianis seem pessimistic. In *La Cecilia*, a group of Italian anarchists settle in the Latin-American jungle, having been granted land there by the Emperor of Brazil. Their attempt to run a self-sufficiency commune along anarchist principles fails because the theorist of the group, an intellectual named Rossi, fails to bridge the gap between his ideas and the working model, the land itself, retreating into a solipsistic reverence for his own grand design, ignoring the problem of dissension which divides the haves from the have-nots within the homestead. Finally, the group faces conscription into the new Republic's Army or deportation to Italy. They do not solve this ticklish problem so much as evade it, singing more marching songs as they stage a scene from *The Death of Danton*.

Where Comolli's film has the edge over the Tavianis' *Scorpio* is in the first-rate script, to which Eduardo de Gregorio contributed. *La Cecilia*, less content with rhetorical gestures (it was made in the mid-Seventies), gains in an authentic analytical focus, showing, not a collapse into frustrated frenzy, but instead the highly plausible conflicts which develop. Comolli's film also gains in subtlety by suggesting the glancing moments when the commune is *almost* working.

Scorpio, conversely, presents a very hostile environment, a terrain of dead lava and potential earthquakes. What both films do signal, however, is the dependence of the respective groups on communal rituals to bolster their ailing morale; both films show the communards singing, dancing and rehearsing theatre spectacles. Both communities stereotype themselves by overreaction.

From the group dream of utopia to that of the solitary activist. This is the shift of focus *San Michele Aveva un Gallo* offers. The film is loosely based on a story by Tolstoy, and the brothers have described the film as a 'symphony in three movements, corresponding to the three seasons of life, youth, maturity and old age'. The hero is Giulio Manieri, who, taking his campaign of anarchist propaganda into Umbria, is met with indifference from the peasants and hostility from the authorities and spends ten years in prison, under sentence of death. During his years in prison Giulio develops a full, not to say florid interior life, projecting scenes from childhood, fantasies, and figments of his own death by drowning onto the walls of the cell. Like Salvatore in *Un Uomo da Bruciare*, he is a man whose messianic vision of political upheaval serves to isolate him from any real awareness of the needs of ordinary people.

In the third part of the film, Giulio is taken away by boat across a lake. His boat passes another, filled with younger socialists with whom he strikes up a dialogue, only to realize that they consider his political ideas out-moded. This shift from the libertarian anarchism of the 1880s to the birth of more organized ideas of socialist collective action is also a prominent theme in Bolognini's *Metello*, set in the same period. Oddly enough, Giulio shares the same death by drowning as Metello's 'out-moded' anarchist father in that movie, though Giulio's death in the lake is self-inflicted. Unable to bear the idea that he is less than politically relevant, he plunges into the water.

The Tavianis have described their hero as being a man who nurtures 'a neurotic impatience with the concept of family ... family appears to him as a form of balance. "It is this balance," he says, "which creates monsters." The only sexual relationship to which Giulio alludes is with a prostitute.' Giulio raises psychological questions which are only developed fully in the Taviani's next film, *Allonsanfan*. In this, the brothers tackle more extensively the difficult questions their previous films have raised, but hedge them rather less behind façades of rhetoric and militant gesture. Fulvio, the hero, is a later model of Giulio, sharing the bourgeois background and the revolutionary fervour. Unlike Giulio, though, he is a man who has lost his ideals and now seeks to promote his own interests, only to find that the residues of his former life as a subversive activist cannot be shed so readily.

It seems plausible to argue that *Allonsanfan* is the first film of mature perception in the Tavianis' repertoire, in the sense that it confronts the complexities of moral ambiguity. *Allonsanfan* is the most intriguing film the Tavianis have made, since it contains a paradox: it is set resoundingly in the past – the period of Restoration following the Napoleonic changes of the 1800s – yet it is cunningly contemporary in its main theme. *Allonsanfan* is a film about the failure of revolutionary commitment. As *San Michele Aveva un Gallo* ends with Giulio's suicide in the lake, so *Allonsanfan* begins, resuming Giulio's story, but with a different line to pursue. Fulvio is released from prison in 1816. As he walks down the steps he is birched by guards. We learn that the Governor has released him in the expectation that Fulvio will lead the authorities to the revolutionaries he once knew as the Divine Brethren.

Fulvio does indeed meet up with the Divine Brethren. As he rounds a corner, he is whisked into a handcart by several Brethren disguised as monks and trundled to the edge of a lake. Unlike Giulio, however, his immersion in the waters is ceremonial not mortal. As the Brethren, under the nominal leadership of Tito, dip Fulvio they interrogate him. Has he betrayed the cause? Fulvio denies that he has. Tito decides to trust him. This absence of guile on Tito's part is characteristic.

The peasant uprising in *Allonsanfan*

Fulvio goes with the Brethren to the house where they expect to meet up with their leader, Philippe. He is dead when they arrive, hanging from a wisteria branch. A suicide note is pinned to his body. Fulvio finds himself unable to react as expected. He wants to go home to find out what has happened to his family. Tito says he will go with him. En route he tries to convince Fulvio that a retreat into the family situation is not realistic. Fulvio's family, being upper-class bourgeois, represents vested interests. Their ideas will inevitably conflict with Fulvio's. They will probably turn him in. Fulvio is not sure. He decides to enter the house disguised as a monk. Tito says he will wait outside for him.

Thus begins Fulvio's realignment with his social roots rather than his acquired political convictions. In the grounds he hears his name being called, by his sister Esther (Laura Betti), by his former nanny, and by his sister-in-law. He realizes that the Fulvio being summoned is his nephew, a handsome child who is playing in the garden.

Fulvio carries off his disguise by saying that he brings news of Fulvio, and joins the family group for dinner. As he retires he decides to test the family feelings towards himself, and announces, in his role as monk, that Fulvio is dead. The reaction of the family is one of stunned grief. Fulvio unmasks; his brother and sister run to embrace him. Fulvio, overwhelmed by intense emotion, collapses.

In the first phase of Fulvio's isolation in the bedroom of his childhood he lies in fever, the walls around him glowing dully golden, in keeping with the subconscious identifications he formed as a child in this room. Fulvio lies still, watching the cat, in a house which has become mysteriously silent (the nanny has even ordered the gardener to muffle the cockerel). At night he passes a candle round the familiar mementoes, glimpsing the books, the violin, and the painting of the *Mayflower* – America still remains a powerful escapist dream for Fulvio, we later realize.

The second phase of the sequence shows the end of fatigue and stasis, and is predominantly white, as his former nanny bathes him in the sheets and notes his erection with a nanny's matter of factness. 'You're getting better, sir,' she tells him. Fulvio is ready to rise and face the deep confusions of his life. 'Make me look nice,' he orders his barber as he absorbs the vista from his balcony.

The final phase of Fulvio's incorporation into his family occurs when he leaves the sick-room for the terrace below, where he sits, wearing his favourite gold scarf, next to his sister. Esther celebrates the 'restoration' of Fulvio by bursting into song – the 'dirindindin' which they used to sing in childhood. This song, which the Tavianis have described as 'full of seductive negativity', is balanced later in the film by the 'saltarello' danced by the Brethren, which is far from seductive but is expressive of more positive emotion. During Esther's song, Charlotte, Fulvio's ex-mistress, arrives. Fulvio goes to embrace her. She has never met his family and is also a member of the Brethren. She insults his family in Hungarian while smiling at them sweetly through the introductions, and as soon as she is alone with Fulvio, makes two demands which he cannot meet. She wants to make love; and she wants him to sell jewels to raise funds for the Brethren. With arms, she says, they can head south where the peasants are already in revolt and conditions thus favour the insurrectionary dreams of the Brethren. Fulvio's plan is more prosaic. He wants them to collect their son Massimiliano, at present being boarded out, and flee *à trois* to America. Charlotte up-ends her supper tray in anger at his weakness. But Fulvio's dream of peaceful escape is already shattered. Through the window he sees the Brethren coming across the grounds, disguised as a hunting party.

At this moment Fulvio's reactions become wholly unequivocal. 'Why have you come to take me away?' he mutters, addressing the Brethren while they remain out of earshot. 'You've been at it for twenty years, chasing ideals that are tired habits. You should have drowned me. I like it here, where everyone knows me. I've lost faith.' One Brother in particular he singles out for his invective, as he ambles into the ambush with the look of a dreamer not a hunter. 'You'll trip,' sneers Fulvio. 'I can't stand the way your eyes always look towards the future.' Esther interrupts him; she has turned in the Brethren. Having overheard his conversation with Charlotte she guessed they would come to collect Fulvio. Faced with a brutal choice – should he warn his former allies or remain in sequestered comfort? – Fulvio vacillates. 'I'll count to ten, then warn them.'

But it is Charlotte who takes the initiative, and runs to warn the Brethren. She is shot down by a soldier. The grounds become a scene of carnage, dogs yapping between the warriors, Charlotte crawling badly wounded towards a carriage. Fulvio and Esther lurk indoors, the curtains drawn, while the battle rages, but finally the sight of Charlotte in physical distress is too much for Fulvio. He helps her into a carriage, grabs his nephew as a hostage and they make their escape, Esther chasing them with a whip.

Once clear, Fulvio releases Fulvio Junior, donating his gold scarf to the child, and he and Charlotte drive to pick up Massimiliano from his foster-parents. There, Charlotte dies. Fulvio and his beautiful six-year-old son linger only to see her buried and then depart with the remains of the Brethren – those who have survived the ambush. With them is a new member, a silent young man with wary, intelligent eyes – the Allonsanfan of the title. He is the son of their former leader Philippe and, we soon discover, made of sterner stuff than his somewhat quixotic co-subversives.

Fulvio tells Tito he will organize an arms deal with the funds entrusted him by Charlotte, and agrees a rendezvous. In fact, he plans to sail to America with Massimiliano. This encounter has taken place at the edge of the cemetery, in a rectangle of foundations. Later, the visual metaphor of foundation stones recurs, in another crucial meeting of the Brethren, to suggest the preliminary nature of the great revolutionary task.

Left alone at last with Massimiliano, Fulvio dines his son in grand manner at a hotel, commandeering a violin from a busker and bursting into a virtuoso solo which leaves the house musicians winded. Fulvio, celebrating his reunion with his son, seems fully alive for once.

As Fulvio carries his son to bed, Massimiliano asks what 'mon amour' means. Fulvio explains. 'Mon amour,' sighs the child, snuggling against him. Fulvio walks to the bedroom window, caught in a strange fullness of emotion. Later that night, however, we see the darker aspect of this love he feels for his son. When the child wakes, sweating in a nightmare and wanting to go home to his foster-parents, Fulvio panics. He falls back on duplicity. This scene is one of the most powerfully rendered in the film. To keep the child in the room he warns him that there is a monstrous toad outside the door. To increase the sinister atmosphere he covers the nightlight in a chiffon scarf, bathing the room in a bizarre green light. When the child, after a mighty internal struggle, resists the idea, Fulvio tightens the screw. He takes away the light and tells Massimiliano that he is right, there is no toad outside the door, the toad is here, on the bed, and has just eaten Fulvio. Only a kiss from a child will bring father back to life. The child succumbs. As he walks to the bed we feel the impact of the trauma, and the film achieves a tragic dimension. Has Fulvio, then, won? There is a knock at the door; as it swings open, a giant toad squats on the threshold, in Fulvio's brief hallu-cination, before a hotel servant enters.

The toad scene, which breaks Massimiliano's will, commits Fulvio to a course of treachery and murder from which there can be no return, as for Macbeth following the murder of Duncan. On a lake he waits with another Brother, Lionello, for gunrunners who do not arrive because they do not exist. On the bank waits a girl, Francesca, who has been enlisted into the movement and who is infatuated with Fulvio. In her innocence, she sees him as a hero. Lionello becomes agitated. Of all the tender-hearts among the Brethren he is the most neurotic. Fulvio proposes gently that they should kill themselves; there is no other way to end in honour. Lionello allows himself to be weighted down with stones and stands on the gunwale. Suddenly he hurls himself back into the boat, declaring that he has understood. Fulvio is testing his nerve to see how strong his revolutionary zeal is. In his excitement he rocks the boat and tips out. Fulvio watches Lionello drown. Francesca runs along the bank, horrified by the scene she has witnessed.

Fulvio lets himself be 'rescued' by a boatload of merrymakers bound for a carnival. The second boat, a deliberate echo of the motif used at the end of *San Michele Aveva un Gallo*, suggests the degree of distance the Tavianis have achieved from their earlier, more solemnly rendered films, for this scene is rendered in the burlesque idiom. In a tavern bedroom Francesca confronts Fulvio, accusing him of letting Lionello drown. Once again Fulvio's manipulative sense is alerted. How can he bring the girl round?

'Yes, I did it,' he begins, 'I let my friend die. Now, if you've got the guts, try to compare his love to mine for you ...' And so, once again, an innocent is snared in the cobweb of false emotion Fulvio weaves so adeptly. For Francesca is hearing what she most wants to hear, that Fulvio murdered a man for love of her, and yet she also senses dimly that what he is telling her is intolerable. 'I hate you,' she says. Fulvio knows he has won her over, 'You won't always be beautiful,' he tells her softly. 'You'll be in gaol for being a revolutionary. You'll be old and childless.'

'If only I could read your mind,' she says. Later that night she takes a willing youth from the carnival into her bed. Having surrendered (sullied) herself, she is Fulvio's. 'What would you like me to do?' she asks him, kneeling naked before him. Later, while she sleeps, he watches her, and this shot dissolves into another very similar composition, catching the Janus-aspect of Fulvio's character. But he has gained an empty victory over Francesca, as over Massimiliano. During the night, his unconscious mind tells him as much; he has another hallucination, in which Lionello surfaces from the waters of the lake, pale and Shelley-like, to reproach him. Fulvio tries to rebut the image. 'Idiot. You died because you couldn't swim.' Lionello replies, with a subtlety the dead man himself could hardly have mustered, 'You're too intelligent. You came too early or too late.'

Fulvio, rattled, decides to leave at once with Francesca. They will stop off at Geneva and travel from there to America. In bed, Francesca proposes a refinement to his plan to elude further commitment to the Brethren. She will shoot him in the leg, she says, doing so before he has time to argue.

The carnival scenes now become foreground material, as merrymakers burst in on the lovers and a large-bosomed lady daubs Fulvio with makeup. Francesca laughs at him for not being able to laugh at himself. Fulvio, like Clerici in *The Conformist*, is a man in whom natural instinct and zest has been replaced by a continuous guardedness. This emerges more forcibly in the scene which occurs a little later, as the Brethren devise their Southern dance; again Fulvio cannot participate emotionally. The Brethren have turned up, not to meet Fulvio, who is no longer important to the project, but to incorporate a peasant rebel known as Vanni, who regales the rebels with the story of how he slaughtered several soldiers who had captured his cholera-stricken wife. She was dead when he got to her.

Cholera is rife in the South, Vanni briefs the Brethren; the people are starving; they have no resistance to the fever. His story moves Allonsanfan, who embraces a true brother in adversity, recognizing in Vanni a fanaticism which matches his own. Tito's plan is that the twenty remaining Brethren should sail with Vanni, using old guns, sporting their scarlet tunics.

The carnival carousers intrude at this moment. A woman is about to give birth, they scream, spreading her on the table. She extrudes a small toy dog, amid gales of laughter. The woman asks Tito why they are lurking in this barn. 'Trying out a new Southern dance,' says Tito. Led by the equally quick-witted Vanni, the Brethren extemporize their brutal 'saltarello', watched by Fulvio, his leg in bandages. Francesca announces that they must take Fulvio with them. Fulvio has passed out; he has no say in his destiny. 'The sea air will do his wound good,' one Brother says drily. So they take him along, as a kind of mascot.

On the boat Fulvio regains consciousness, believing for one exultant moment that he is bound for America. The reality of his situation enrages him. Tito calms him, telling the Brethren that he has been making too many of the decisions. Perhaps he pursues the revolutionary cause for egocentric reasons. For all that, he adds, in a world where almost everyone seems asleep, it is difficult to curb one's passion for violent change. Tito's introspection is interrupted when land is sighted. Allonsanfan is the first to don his tunic. As night falls, and the Brethren sing the *Internationale*, Fulvio interrogates Vanni concerning the peasants' reaction towards him in his home region. He learns that Vanni's nickname is 'Pesto' (Plague), thus isolating the potentially inflammable elements in the present situation which he can turn to his advantage.

While the Brethren march inland crying, 'We are for the people,' and finding precious little response among the deserted fields, Fulvio slips off, to tip off the local priest that twenty armed men have landed on the island, along with Vanni Pesto. The linking of the rebels with the local 'monster' serves its purpose. The priest retorts glibly. Fulvio leaves the priest preaching a sermon which incites his 'sheep' to become wolves and tear the rebels to pieces.

Again, the Brethren have convened in the foundations of a church. Fulvio announces that he has spoken to local guerrilla leaders, who will join the cause and overthrow the tyrants. Fulvio's arrangement with the priest guarantees him physical protection once the massacre begins. He alone will be without the scarlet tunic. This ruse, however, contributes to his own death. As the

Brethren line up in the fields to greet the advancing horde of peasants Allonsanfan is the first to realize what is about to happen. The Brethren panic. Vanni fires the first shot, wounding a child. Tito then shoots Vanni, immediately running forward to embrace him. All but Allonsanfan are then clubbed and pitchforked to death. Allonsanfan, fleeing, catches up with Fulvio.

What follows is the most ironic scene in the film. Allonsanfan, whom one British reviewer dismissed as 'playing no significant part in the action', confronts Fulvio with a Judas stroke of his own. 'The peasants have joined our cause,' he announces jubilantly. 'They all joined hands and danced the *saltarello*. A little girl was laughing. Everyone clapped. They embraced one another.' Over Allonsanfan's peroration we see this vision of utopia enacted. Fulvio, who has everything to lose by accepting this story as reality, is finally deceived when Allonsanfan, in a master stroke, removes his red jacket and runs on. Fulvio dons the jacket. Soldiers appear. One fires, and kills him. Bells ring out across the land. The power of the status quo prevails. Yet Allonsanfan has escaped, in Fulvio's white shirt, to fight another battle. The revolution has been lost in deed but not in vision. Allonsanfan's utopian dream has been vindicated with the just execution of Fulvio, the revolution's betrayer.

In *Allonsanfan*, the Tavianis finally achieve the formal control within their cinema of 'utopian research' for which they have been striving in the cycle

Marcello Mastroianni in *Allonsanfan*

of political films which precede it. The film is a triumph of pure cinema, in which music, *mise en scène*, cinematography and a script of satisfying psychological density conjoin with considerable emotional and intellectual tension. Talking to Jean Gili in 1975, they commented, a shade wrily, 'From a masochistic point of view, Fulvio is a part of ourselves. It seems to us today that in making this film we have liberated ourselves.' *Allonsanfan*, superseding the rhetorical tendencies of the earlier movies, incorporating within its essential structure the dilemma of the film-maker who, however much he wishes himself into a revolutionary role, is finally a member of the bourgeois intelligentsia, marks the moment in the Tavianis' career when they become able to dramatize their own contradictions. It is perhaps significant that only in *Allonsanfan* does the utopian vision hold intact, and the ending, while it is all too plausibly brutal in reality, remains emotionally optimistic. With their portrayal of Fulvio, a man in touch with his bad streak, they remind us of the value of the aphorism, 'The strength of vice is that it refuses to tolerate mediocrity.' The Tavianis' two most recent films, however, serve to consolidate the gains achieved in *Allonsanfan*, rather than transcend them.

Padre Padrone, of 1977, took the Cannes Critics' Prize of that year and received mighty accolades from *cinéastes* as diverse as Rossellini and Wenders. The film is based on Gavino Ledda's book about his own experiences as a shepherd boy in the remote hills of Sardinia, who goes away to the mainland to serve his time in the Army, and thus breaks the cultural and emotional isolation which has been imposed on him since early childhood, partly by stringent economic conditions, partly by his patriarchal father, Efisio. While in the Army, Gavino begins to make friends, and concentrates on his studies. He becomes a graduate (later a Professor) in linguistics, and then returns to Sardinia to resume the conflict with his father where he left off. Now, however, he has the cutting edge and his father has to acknowledge it.

There are two scenes in *Padre Padrone* which convey the film's essential purpose. The first occurs when Gavino constructs a radio, during his period of Army training, and finds it works, thus understanding that he, too, can become an achiever and not merely a passive victim of circumstance. The second occurs when he finally defeats his father, and all those oppressive values of domination and acquisitiveness which he sees Efisio as representing. In his moment of defeat, Efisio hears the thunderous gong which is the acoustic symbol the Tavianis have established to represent the crushing silence of Gavino's childhood solitude in the hills.

It is plain to see why Ledda's account should have appealed to the Tavianis. Not only his main theme – self-amelioration through the auto-didactic path – has always been theirs, but also the specific dynamics of the narrative correspond with their own screenplays – the search for the mainland where more utopian conditions might prevail, the misery of peasant life, and above all, their most grandiose theme, the death of the father, which

Saverio Marconi in *Padre Padrone*

played so prominent a part in their *Sovversivi*, hinged as this film was around the death of Togliatti. To English audiences, the auto-didactic triumph may seem a somewhat revanchist theme – one which achieved its pinnacle of expression in British culture around the turn of the century; in Hardy's *Jude the Obscure*, for example, the theme achieved its most tragic dimension. It should be remembered, however, that Italy created its democratic constitution very much later, and then suffered a severe setback to democratic development during the era of Fascism.

Gavino's decision – to return to Sardinia in order to pass on what he has learnt to his inflexible father – might seem quixotic, but is in line with similar decisions which the Tavianis themself have imposed on their earlier (fictional) heroes. If he had stayed on in the city, Gavino has said, he would have become a recluse in a cave. In Sardinia, his sense of rootedness serves him as a kind of scourge, by which he can muster a sense of dynamism and mission. His achievement, he claims, is not so much a personal triumph as a distillation of the accrued rebelliousness of centuries of solitary shepherds. The film ends showing Gavino, wearing symbolic red, rocking himself in the village, as he used to do when a child. The last shot, though, shows children in the classroom, and the sound of the accordion – an instrument Gavino learnt to play

213

while living on the hills. The full-circle is, again, not merely pessimistic, but represents Mao's consciousness-raising circle, as the Tavianis would have it. This is made quite clear in the soaring Strauss waltz which occurs (as a reprise) on the soundtrack.

The use of sound, as always in the cinema of the Tavianis (whose first scripts were produced on the radio before they became fiction film-makers), is striking throughout the film. Subjective sound replaces dialogue, as it did also in *Sotto il Segno dello Scorpione*. The bleating of sheep and goats, the rushed breathing of the shepherd boys masturbating or bestializing their live-stock, the clenched climaxes of their parents in family bedrooms, the jubilant strains of the waltz when Gavino first plays his accordion, and the strains of Mozart from Gavino's radio; even in the city, certain atavistic sound memories are carried by Gavino, who hears in his mind the cawing of rooks and the bleating of lambs.

There is one scene especially which suggests the paramount importance of sound in the film's meaning. This occurs when the young shepherds stage a singsong to celebrate their imminent escape from their hated island to the industrial cities of Germany. They sing *bierkeller* songs while their fathers, who are staying behind, compete with hymn chants beneath the same canopies. The opposing values are conveyed through sound rather then image. Elsewhere, however, the imagery of father-opposition comes first: as when Gavino leaves the island in a truck, along with other potential con-scripts, and makes a gesture of his contempt for the island, and his father, by urinating on the road from the tailgate of the truck, cheered on by his peers.

It is possible to argue that the Tavianis, with this film, have not made the vital gains that one might have expected after *Allonsanfan*. The film has been almost universally acclaimed, yet only one reviewer has isolated its weak-nesses. Gavin Millar's notice in *The Listener* seems so apposite that it is worth citing in some detail: 'All the time,' he comments, 'there is the intruding suspicion that an intelligence and a sensibility more calculating, less open-minded than Gavino himself is manipulating both him and us towards a didactic end that, left alone, we might resist. Gavino is not so much observed as made an example of. He even becomes, at times, a piece of landscape, or scenic design. When the boy guarding the neighbouring flock takes him under his wing, he does so with a rhetorically literal metaphor, sweeping his great yellow blanket suddenly round Gavino's shoulders ... The image seems meant to impress rather than to touch, and the gesture is out of scale with any reality you could imagine ... Or perhaps any reality we could imagine. The English are not great on epics.'

These perceptions are useful in relation not only to *Padre Padrone* but to the Tavianis' form of political cinema in general, which combines a somewhat lofty intellectuality with a visceral coarseness deriving from their attraction towards Neapolitan culture, in particular burlesque and opera. What seems missing is the region *between* head and gut, the emotional delicacy which enables them to transform rhetoric into sympathetic identification with their characters outside the utopian arena and inside life as lived. In *Allonsanfan*, through the characterization of Fulvio, they are able to treat psychologically this emotional hollowness, make it the central tension of the film, as Fellini made his contempt for Casanova the pivotal tension in that movie. The stridency of the father-son conflict in *Padre Padrone* does seem to signal an emotional vacuum; where the film should be most densely and dispassionately observed, it becomes oratorical. Bearing in mind the similarly couched conflict between Amerigo and his father in Bolognini's *La Viaccia*, one could also say that this element is somewhat derivative, adding nothing to our insights into agrarian family rivalries which hasn't been realized much earlier in Italian cinema.

What is interesting about *Padre Padrone*, however, in relation to the Tavianis' earlier films about the utopian goal, is that, in the process of their passionate identification with Ledda's theme, they are obliged to flesh out the utopian specifications of their earlier work within the framework of a concrete social organization – Italian society itself, and the institutions of school, family, college and Army.

Il Prato links in theme with *Padre Padrone*, in the sense that its three protagonists, Giovanni, Eugenia and Enzo, in effect develop Gavino Ledda's notion of a contemporary return to the land, suggesting that at last the

Omero Antonutti (L) and Saverio Marconi in *Padre Padrone*

Padre Padrone

Tavianis are resolved to cast their ideas of agrarian utopias within a firmly contemporary mould.

For Enzo, horizons widen when he meets Eugenia (played by Isabella Rossellini) and they commandeer an empty villa, where Eugenia stages her 'meadow theatre' – using the local landscape of Tuscany as the setting for her children's fable, helped by local children. The problems of self-determination are defined by the arrival of Giovanni, a law graduate who has defected from law to pursue his real love, cinema. All three young people are shown to be in defiant retreat from bureaucratic and commercial values. Eugenia, too, is on the run from a mundane job in a tax office in Florence. In this respect they bear some affinity with the protagonists of Alain Tanner's *Jonah, Who Will Be 25 in the Year 2000*.

Giovanni's rapport with Eugenia is triggered by his father's decision to send Giovanni to the empty villa in San Gimingano, owned by him, with two rooms left to sell. Giovanni forgets commerce when he meets Eugenia, out in the meadows teaching children to walk on stilts. Eugenia, while she accepts Giovanni's love, is unable to break off her attachment to Enzo. The latter accepts this uneasy *ménage à trois*, but Giovanni suffers acutely, writing to an old friend, 'None of the three of us think a man or a woman must love only one person in their lives. Until a month ago I'd have laughed at anyone who felt that way. But now that he's here I must leave. Because he detests my friendship.'

Enzo returns to Milan and attempts to immerse himself in his legal career, but learns that Eugenia is about to leave for Algeria to teach in a community of Italian children. Enzo, who has been forced to abandon the squatted villa, is going with her. Giovanni meets Eugenia once more, but she tells him she cannot stay. Giovanni kills himself in the villa, telling his father on his death-bed, 'I ask only one thing. Tell Eugenia.'

The somewhat mawkish ending to *Il Prato* does help to explain why, hitherto, the Tavianis have avoided the realistic story, set in the contemporary world; now, in handling such terrain, they reveal a somewhat novelettish tendency – the obverse, perhaps, of their earlier predisposition to rhetoric. The need to make their dramas *perform* suggests an uncertainty with the emotional elements of their cinema which is very different from the reticent clarity of Olmi's world. Yet, if *Il Prato* remains a less than fully resolved incursion into the modern world and its attendant conflicts and identity problems, it does at least touch upon crucial experiments within modern life – in particular the search for alternatives in the 'new peasantry', the young dissidents of the technological society, with their yearnings for self-expressive ways of life, and their frail hopes of Arcadias around the corner.

7. Terminus and Tantrum: Luchino Visconti and Lina Wertmuller

In his monograph on Visconti (published by Secker & Warburg in 1967) G. Nowell-Smith concludes by commenting that Visconti's career in cinema, which began in France, and ended when he died in 1976, is characterized by an opposition between two imperatives: nostalgia – his emotional roots in the past, and progress – his belief in the value of projecting intellectually into the future. It is interesting to compare this view of Visconti's cinema with Visconti's own ideas, as expressed in an article entitled *Cinema Antropomorfico* (published in *Cinema* 173/4, September/October 1943 and first published in English in the collection of essays *Springtime in Italy*, Talisman Books 1978). What Visconti meant by an 'anthropomorphic cinema' he defines in the following comment: 'My experience has taught me that ... the human being acquires truth and character thanks to the emotions he undergoes, while his temporary absence from the screen will cause things to return to the state of non-animated nature ... reality is made by man and continually unmade by him.'

It is clear, then, that Visconti, writing at the time he was making his first feature film, *Ossessione* (1943–5), was a culturally integrated artist, with a strong belief in the value of the human presence in cinema, and an equally stalwart faith in emotional as opposed to purely theoretical socialism – he also remarked, in the same essay, 'A human being is always at bottom, "freeable" and "re-educatable".'

The integrity of the individual, in life as in cinema, was a prime belief in Visconti's *La Terra Trema* of 1947, never more so than in the moment when Antonio, having failed to make the fishing cooperative work, cries, 'I did what I did for everybody, not just for myself. And now? You see how they've all walked out on me ... We have to learn to stick up for each other, to stick together. Then we can go forward.' Antonio's sense at this moment of the failure of his socialist experiment is central to the film's statement, yet not quite in the manner most critics have defined it. Antonio's

Luchino Visconti directing *The Damned*

speech does not so much locate *La Terra Trema* in the heart of the neo-realist experiment of the middle 1940s as sound the death-knell of the political hopes associated with the euphoria of Liberation.

The social desperation of the fishermen of Sicily, as focused in Visconti's second movie, has become more fragmented by 1960, with *Rocco e i suoi Fratelli* (*Rocco and His Brothers*). Once again, the problem of unemployment in the South is the central social theme. This time, however, the setting is Milan, in the boom years of the late Fifties. The Pafundi family migrate to Milan from the South. One son, Vincenzo, has already settled in the suburbs, found a job, married and become a Northern industrial worker. Rosaria, the matriarch, intends to get her other three adult sons working. They are Simone, Rocco and Ciro. All three begin by shovelling snow, which they have never seen in the South. Simone becomes a boxer. He falls for a whore (Nadia). She sticks with him for his money. Later, she and Rocco become emotionally involved. Simone takes his revenge, raping Nadia and assaulting Rocco. This leads to the strangest scene in the film, which Visconti underlines by setting the encounter on the roof of Milan Cathedral. Rocco tells Nadia, in effect, that his impulse to protect Simone from his own violence is stronger than his feelings for her. In rejecting Nadia, he exposes her to further ferocity from Simone. Simone waylays her in a lonely riverside haunt and stabs her to death, arriving home dazed and bloody. Rocco, again, seeks to protect Simone, this time from the law. But Ciro, the younger, more pragmatic brother, has had enough. He turns in Simone. In the final scene Ciro meets Luca, the youngest of the brothers (who is still of school age), outside the motor factory where he works. Luca has half an idea that he would like to go back to the South. Ciro has no such illusions. He is the first member of the family fully to integrate in the Northern industrial mould.

In other words, in these two films Visconti examines the hypothesis of socialist altruism – Simone becomes a boxer not because he wants money and fame himself but because he wants to provide for the family, just as Antonio attempted to establish a fishing cooperative – and shows how it seems doomed to fail. This is an important motif to isolate, not least because it contravenes the idea among European critics that Visconti, once a self-professed Marxist, always remained so. In fact, as Monica Stirling points out in her book, *A Screen of Time* (Secker & Warburg 1979), Visconti in later life was essentially a liberal democrat, hence his admiration for politicians like John Kennedy.

Visconti's cinema, even more than his parallel career in theatre (over forty major theatre productions and around half that number in the field of opera), has always 'swung between the old world and the new', as Prince Salina would put it. He has shown the poor and deprived ('the broken ones' in Verga's phrase) in *La Terra Trema*, the deracinated and beleaguered migrants in Milan (*Rocco and his Brothers*), the petty-bourgeois erotic desperations of *Ossessione*, but also the grandiose decadence of the dying European castes – the landed nobility of Sicily in *The Leopard*, the doomed, homicidal industrial Nazis of *The Damned*, the demise of the intellectual Elites, *Death in Venice* and *Conversation Piece*, and the baroque morbidity of Nietzschean monarchs and nobles, *Ludwig* and *The Innocent*.

Comparing Visconti's cinema to Olmi's, say, or Antonioni's, there is not the closely observed detail of daily life; this sort of visual and psychological acuity escaped Visconti. Visconti became an institution in Italy precisely because he was not original. His great strengths were his obsessional concentration on the surface detail – of costume, set, colour tones, actors' gestures – and of course, his own cultural refinement, which enabled him to compensate for his absence of personal vision by exercising a fastidious electicism towards all his activities in the arts, from opera and theatre to cinema and, latterly, television drama productions.

In line with this, it should be added that Visconti's absence of personal vision can be seen in the fact that he rarely wrote his screenplays. His regular writers were Suso D'Amico and (latterly) Enrico Medioli (a distinguished novelist in his own right). Few of his films were written directly for the screen, but adapted from books, plays or stories.

Before we attempt a discussion of Visconti's three major films of the Seventies, let us briefly consider his most challenging creative task of the previous decade, his screen adaptation of Lampedusa's *The Leopard*, one of the major novels of the century (some would argue, *the* novel). *The*

Alain Delon and Claudia Cardinale in *The Leopard*

Leopard raises so many themes and motifs which are integral to Visconti's work, particularly as it has presented itself to us in the final three films, that some delineation seems desirable.

It is superficially tempting to relate Prince Salina, the hero of *The Leopard*, to Visconti himself, for Visconti too belonged to the Italian aristocratic class, being the fourth child of the Duke of Modrone. Visconti's biographer, Monica Stirling, believes that this accident of birth was crucial to Visconti's subsequent development as an artist and cultural *padrone* in Italy's performing arts community of post-war years and onward, commenting: '... A child like Luchino Visconti, with ancestors still moving in tapestries all around him, is born with a sense of history's continuity into a world of vast imaginative

221

dimensions, full of communicative ghosts with familiar faces.'

Perhaps, on the other hand, one could argue that all Italy's major film artists belong to a tiny cultural, if not social, elite in a nation where the cinema has been so central to the collective sense of cultural dignity. And what Ms Stirling attributes to Visconti – that sense of history's continuity – is even more a component of Lampedusa, who understood that nothing in society ever really changes, since all forms of political, bureaucratic and technological stimulation can only touch the surface of things, leaving the lot of the ordinary working person unaltered.

If quietism is Prince Salina's central unspoken philosophy, for Visconti, a Northerner with none of Lampedusa's natural fatalism of outlook, Sicilian apathy would have seemed a problem perhaps, rather than a solution. This is exactly the issue which confronts Carlo Levi in *Eboli*. As a man of the North, he perceives the 'backwardness' of Southern Italy as a feudal anachronism. But when he becomes a resident of the South (albeit a re-luctant one) his central insight is Lampedusa's – there is no greater social problem here, just a different form of confusion.

What Lampedusa specifies, through his hero the Prince, is nothing less than the non-imperialist personality, the man 'with a disposition to seek a shape for life from within himself' as he expresses it elegantly. This kind of self-integration is what we note, too, in Carlo Levi, and for that matter in Olmi's peasants in Bergamo. Indeed, Prince Salina, like the peasant Anselmo in *The Tree of the Wooden Clogs*, is a self-stimulating, self-organizing man. Salina studies the stars as Anselmo invents stories and persuades his tomatoes to fruit early. Visconti, however, with his inveterate interest in the operatic elements of his protagonists' destinies, seeks to underline the 'heroic' aspects of Salina; the implications of the integrated as opposed to the conflicted per-sonality are less accessible to him than to Lampedusa.

A comparison of Lampedusa's text with Visconti's film sets out Visconti's limits. Above all, Visconti's film ignores the vital rhythm of the book, which is constructed in eight passages. The final section is dated 1910, 27 years after the Prince's death, and is titled Relics. It shows the breaking up of the Salina household's religious relics by an efficient demolitions expert from the Vatican. David Nolan, in his essay on the novel in *Irish Studies* (Winter 1966), has aptly described the novel's formal control: 'The external looseness of construction is deceptive, for Lampedusa has selected an exact minimum of scenes. The book is not an anthology, but a seamless whole . . . That his scenes are all effective and necessary may easily be seen by attempting to remove any single one of them.' Visconti's film, however, proposes startling modifications to this 'seamless whole'; and, in so doing, alters the purpose of the novel, making a classically composed celebration of manners into an epic based around the heroic protagonist, Salina.

The historical context in which the Salina relics are destroyed is one of the Restoration, in which the recent social upheavals of Garibaldi's landing and the political restructuring which followed it are consolidated by the new dominant class, the bourgeoisie – 'instead of a class they become a "caste" with specific cultural characteristics, but no longer with predominant economic functions.' It is this shift – acutely defined by Gramsci in his essay, *Notes on Italian History*, of 1935 – which Lampedusa's final, vital section symbolizes. Visconti, failing to incorporate this important section, appears to do less than justice to his professed sensitivity towards Gramscian concepts of Italian social organization, let alone Lampedusa.

It could be said, of course, that since Twentieth Century Fox changed Visconti's film to the point where he could no longer associate his name with the project, any criticism based on the existing British print of the film is inevitably provisional. On the other hand, Visconti evinces a similar somewhat peremptory attitude towards later adaptations of classic works. Let us consider briefly Visconti's other adaptations of major literary texts – 'novels people are pious about,' as Michael Wood expressed it – *Lo Straniero* (*The Outsider*) and *La Morte a Venezia* (*Death in Venice*).

Visconti was offered the adaptation of Albert Camus's novel by Dino De Laurentis. He accepted on the strength of his admiration for Camus' work but soon found himself directing a film in which 'he could not truly participate', as Monica Stirling has commented. He disagreed with De Laurentis about the form the script should take – and he failed to obtain Alain Delon for the main role.

Committed to the film by agreement nevertheless, Visconti executed the project with his customary concern for cosmetic detail, ensuring, for example, that bathing huts and costumes of precisely the correct vintage were installed on the beach where he shot the footage – the film was set, like the novel, in the Algeria of 1938. Visconti was criticized widely upon the film's release for adhering scrupulously to the surface detail while failing to explain adequately the twisted psychology of the hero Meersault (Marcello Mastroianni). Once again, Visconti attributed the responsibility for the film's imperfections to the insensitivity of his producers.

With *Death in Venice*, of 1970/1, Visconti had the support of a creative producer named Mario Gallo. Visconti adapted Thomas Mann's novella in collaboration with his regular co-scenarists and managed to obtain the services of two artists he much admired, Dirk Bogarde and Silvana Mangano, for the two major roles. The ostensible story of the novella is simple: an elderly German composer (based in broad outline on Gustav Mahler) goes to Venice for a long rest and becomes infatuated with a fourteen-year-old Polish boy, Tadzio, who is staying *en famille* in the same hotel, along with his beautiful mother, three sisters and the governess/nanny. Von Aschenbach's infatuation with Tadzio takes him over; he tries to leave, fails, becomes delirious as a consequence of his emotional agitation as well as the cholera which is invading the resort, and finally collapses and dies on the beach, while watching Tadzio wade into the sea.

British critics did not much care for *Death in Venice* at the time of its release in March 1971. John Coleman, writing in *New Statesman* (5th March

Silvana Mangano in *Death in Venice*

1971), was disconcerted by the film's imbalance between cosmetic perfection and psychological crudity – 'What worries me ... is whether Visconti's confidence in what his characters have on their backs, in their pockets or handbags is matched by an equal knowledge of what they carry in their heads.' While Michael Wood, writing in *New Society* of 11th March, considered that '... I had thought, as Bogarde tiptoes among the bathing huts ... that the film was almost turning into a high-class version of *M. Hulot's Holiday ...*'

Sound as these judgements may be, they do not locate one essential problem Visconti and his adaptors were faced with in transposing the book to the screen. Mann's novella begins, as Lampedusa's novel ends, with a sequence that is separate from the main body of the plot yet integral to the meaning of the work as a whole. In this passage von Aschenbach is seen in his home town, Munich, in a state of near nervous exhaustion as a result of his intellectual labours. He takes a walk, and encounters an exotic figure with 'a bold and domineering, even ruthless air' which, when he meets von Aschenbach's gaze, becomes 'menacing', thereby provoking in the over-taxed author a sudden urge to travel south.

The significance of this emblematic figure for Mann becomes apparent when, in a scene at the hotel in Venice, von Aschenbach is accosted by a brutal-looking street musician. This is during his period of near-delirious infatuation with the boy, and the face of the man seems the same as that of the red-haired stranger in Munich. What Mann is conveying, in effect, is a *polarization* of von Aschenbach's moral values as he nears his own death, during which phase of heightened consciousness he tends to see humanity in absolute terms – Tadzio as an innocent angel, but these two street vagabonds as diabolical interventions.

Visconti was aware of the symbolic significance of the red-haired stranger and tried to incorporate the opening passage of the novella into the film, using first a chronological approach, then flashbacks. In the end, he rejected the preliminary passage entirely, saying, 'Just the sight of the red-haired stranger isn't enough, doesn't automatically explain von Aschenbach's sudden impulse to travel.' This remark suggests that Visconti's identification with the literary source material, on this occasion as elsewhere, is less than fully empathetic. Without the inclusion of this crucial signal of von Aschenbach's already encroaching delirium, the worship he bestows upon Tadzio does indeed seem, as Jan Dawson put it, little more than 'the self-discovery of a closet queen,' (*Listener*, 11th March 1971).

It would seem then, on the evidence of Visconti's various experiments with bringing major novels to the screen, that his concern for costume and

Dirk Bogarde in *Death in Venice*

225

decor detail (down to the last bus ticket and bathroom tile) can mask an absence of rapport with the underlying patterns of his chosen text.

Before we turn to a discussion of Visconti's last three movies let us mention briefly the two other major films of the Sixties, both of which were concerned with that perennially fascinating theme of Italian directors, the Nazi era, and both of which were original screenplays, though *Vaghe Stelle dell'Orsa* (*Bright stars of the Bear*), of 1963, was inspired by the Orestes/Elektra myth, and *The Damned* (1969) was inspired by certain passages of Thomas Mann's *Buddenbrooks* novel sequence, while Visconti also drew extensively upon William Shirer's *The Rise and Fall of the Third Reich* during his script preparation.

Both films chart the degradation of distinguished bourgeois families, the first in Italy, the second in Germany. *Bright Stars of the Bear*, set in the present, is pivotted around crucial events which took place within a family during the Fascist era, in particular the deportation of the father, a Jewish professor, to a concentration camp where he died, perhaps betrayed by his wife. This is mystery in the family's past which still torments the now adult children, Gianni (or Orestes, as Richard Roud would have it) and Sandra (Elektra).

The trauma of the past intervenes between brother and sister, with ultimately tragic results. During a visit Sandra makes to the family home at Volterra, in Tuscany, Gianni learns that she does not reciprocate his incestuous passion; Sandra is now married, and her American husband comes first. As the memorial service to the father takes place in the grounds, Gianni kills himself; his death is discovered just as the monument to his dead father is unveiled outside.

With *The Damned*, made six years later, Visconti returns to the same thematic terrain, but now examines the decline of a family of German industrialists, the Essenbecks, during the formative years of Hitler's Chancellorship, 1933–4. Authentic events, such as the massacre of Rohm's SA members by Hitler's SS at the Wiessee resort in June 1934, are included, mingling with the sexual perversions and power afflictions of the Essenbeck family, as they align with a treacherous industrialist. Once again, the family is viewed by Visconti as the breeding ground of personal and social neurosis, a motif which recurs in his cinema. A specific psychological refrain which runs through both these films, and will be repeated again in *Conversation Piece*, is that of the mother as termagant, providing the weaker members of the family (husband, sons and lovers) with the false elation out of which they temporarily find the fury to commit atrocities.

In *Ludwig*, of 1972, Visconti goes further back in time, to show the cultural antecedents to Nazism, when the ethos of a regenerated Germany was being established, in the writings of Nietzsche, the music of Wagner, and the grandiose schemes of social and architectural design implemented by Ludwig II of Bavaria, who reigned from 1864 to 1886. Ludwig II, described by Wilfred Blunt in his study *The Dream King* (Penguin 1973) as 'one of the

Ingrid Thulin and Dirk Bogarde in *The Damned*

greatest builders in an age of historical revival', was an obvious choice of subject for Visconti to portray on film, for like Visconti himself he 'demanded before all else historical truth based on his own scientific and literary studies'.

Both the film versions of Ludwig's reign made in the Seventies, Visconti's of 1972 and Syberberg's of 1973/4, are highly selective historical portraits, depicting the grandiosity of his regime – his castle-building and his love and patronage of the music of Wagner – rather than his intensely negative characteristics; he wished, as Wilfred Blunt points out, to rule as an autocrat, and tried to create a monarchy of absolute power along the model of the Bourbon Kings of France. There are, however, considerable differences of emphasis, historically and psychologically speaking, in the two film portraits, and these it might be useful to examine.

Syberberg's film consists of a series of barely animated *tableaux vivants*, in which he displays not only Ludwig's decadent romanticism but also his relationship to the court and fiscal bureaucracies surrounding him. In general, Syberberg's lengthy film shows a more abstract Ludwig, only tenuously attached to the grid of familial and dynastic ties, but connected instead to a changing cortège of young hangers-on, rather like a street apache in an early Fassbinder movie. Perhaps inevitably, both films quote Ludwig's most famous remark that he wished to remain an enigma. Visconti's film, however, makes the monarch less enigmatic. Instead of the detached, pageant-like tableaux of Syberberg he designs elaborate set-piece scenes in which his usual emphasis on costume and scenographic detail is paramount, as indeed is his carefully contrived use of colour – blue is the predominant tone employed, particularly in the exterior scenes set in the castles and snowy forests. Visconti's greater interest in familial relations is also evident, as is his depiction of Ludwig's relations with the Wagners. Andrew Tudor, reviewing the film in *New Society* (26th October 1978), points out that the British version of *Ludwig* is shorn of fifty minutes of material. His conclusion is that 'it is as if, in recutting the film, Visconti has plumped for his fascination with Ludwig at the expense of his ambition to express the forces of history.'

The fact that Syberberg's film, while it is certainly more distantly perceived, and reduces the attachment to chronological and dynastically focussed storytelling, is *also* highly self-referring, suggests that it was Ludwig himself who represented 'the forces of history'.

We have said that Visconti's film is more concerned with tracing the family tensions and interdependencies. This becomes most arrestingly documented in Ludwig's relationship with his cousin, the Empress Elizabeth of Austria (played by Romy Schneider with great panache, as though this favourite actress of Visconti's has finally found a role which suits her talents). Of those around Ludwig, his father, mother, wife Sophie and brother Otto, only Elizabeth is revealed as having much insight into the conflict imposed upon him by his essentially introspective nature and the pressures of his role as monarch.

Despite her fond feelings for her cousin, Elizabeth manages to maintain her dispassionate tone throughout the scenes evoking their troubled relationship. When Ludwig, for example, meets her at his new castle to tell of the triumph of Wagner's latest opera, which he sponsored, she demands bluntly, 'How much did it cost? Triumphs are forgotten. What's your ambition? Wagner gives you the illusion of having created something, just as I give you the illusion of love. You need help I cannot give you.'

Equally blunt with Ludwig, in Visconti's version, is Minister Durkheim, a less shadowy figure than in Syberberg's film. He tells Ludwig, 'You want to increase your freedom to follow your instincts. This privileged liberty is not freedom. You must find a humble mediocre role. You have to accept mediocrity if you have sublime ideals. It's the only way of avoiding a filthy solution.' The realism of such an observation seems to tell us more about Visconti's and Medioli's viewpoint towards the end of the twentieth century than the Minister's.

Visconti shows Ludwig's relationship with his brother. Otto, returning from the war with Prussia, confronts Ludwig in his newly created 'blue room' under a ceiling showing the phases of the moon. Otto says that Bavaria is losing the war. To Ludwig, war is not diverting, though he does defer to Otto's serious mood by switching off the ceiling. 'As far as I'm concerned,' he says, 'the war is incestuous and fratricidal. It does not exist.' Otto, less protected by such magnificent solipsism, returns to the front. When he returns, following Bavaria's defeat, he is seriously depressed; later, his sanity collapses.

For each scene showing Ludwig's grandiosity – the castles of Herren-cheimsee, Neuschwanstein and Lindenhof, the operas of Wagner staged at the specially built theatre in Munich (paid for by Ludwig from the State coffers), the moonlit sleigh rides in the forest, the terse trysts with handsome young soldiers by the lakes or in the hunting lodges – Visconti counterbalances scenes in which matters of housekeeping arise – the Exchequer objecting to more funds for the opera, Sophie's plaints that her husband never visits her, and later the plotting of ministers to depose Ludwig and make his uncle Leopold the Regent. Thus, Visconti constructs his film so that a tension exists between Ludwig's own vision of his destiny as 'Dream King' and the growing disaffection of those in positions of power and obligation around him. What is evident about Ludwig, in Visconti's film, is that he was a man intent above all on creating monuments to his own values, who died, significantly, before the master work, his sumptuous throne room at Lindenhof, could be completed. In Visconti's film, Ludwig's death remains enigmatic.

During the four films he made during the Seventies, Visconti showed himself to be drawn to terminal themes – old age, solitude, rage, the preparation for death, and the struggle against the irritations of physical debilitations. In addition, all four heroes, von Aschenbach, Ludwig, The Professor of *Conversation Piece* and Count Tullio of *The Innocent*, are men whose

Helmut Berger and Romy Schneider in *Ludwig*

fundamental characteristic is their extreme isolation from conventional social groupings or categories.

Gruppo di famiglia in un interno (*Conversation Piece*), adapted by Visconti and Medioli from a novel by Medioli, was realized in 1974, two years after Visconti had suffered a serious stroke, and its form was dictated by his physical fragility – he directed the film from his wheelchair. The film is set entirely indoors, in the apartment of the hero, known simply as The Professor (Burt Lancaster), and in the apartment above his. It is a 'chamber' film, concerning an elderly intellectual unable to establish personal relationships who isolates himself from professional and social roles in order to think and study. His particular interest is painting; he collects 'conversation pieces' of the eighteenth century. Visconti himself, in an interview in *Avant Scène du Cinéma* (No 159, June 1975), has described The Professor as 'an egoist who substitutes objects for relationships. He has a collecting mania. What counts is men and their problems.'

It is The Professor's 'collecting mania' which proves his Achilles' Heel, as we see in the opening sequence, when he toys with the idea of buying a Blanchard landscape offered by a visiting dealer. While he debates his housekeeper admits a wealthy woman, Bianca Brumonti (Silvana Mangano), her son, daughter and her lover Konrad (Helmut Berger). They want to lease The Professor's upstairs apartment, which is standing empty. The Professor has no intention of letting, however, though he does allow himself to be chivvied into showing round Bianca and her brood. The argument is curtailed when Bianca learns that the police are towing away her car. The Professor, having also decided not to buy the painting, retires with relief to his solitude. Later, however, when he phones the dealer to say that he wants to buy the Blanchard after all, he learns that is has been sold. The buyer is Bianca Brumonti.

Thus The Professor's weakness, his acquisitive urge, has led him into a trap, for Bianca now reappears to present him with the painting, and to renew her assault. 'I'm an old man,' he protests weakly. 'Hysterical. I can't rent the apartment.' But he does. And before long all his initial fears are confirmed. The Brumonti family and Konrad begin renovations as soon as they move in. While The Professor is out one day, the living room ceiling collapses. The Professor, returning, finds only Konrad in residence, who apparently takes his side against Bianca and urges him to make her 'pay millions' for the damage caused – her husband is a right-wing tycoon.

In these early scenes two perennial themes in Visconti's cinema are incisively established: the awfulness of the new rich bourgeoisie at home (a theme Visconti treated with relish in his malicious episode-film, *Il Lavoro* in 1962), and the latent eroticism of the old man's rapport with the young man, which echoes similar relationships in his previous films, Salina/ Tancredi, von Aschenbach/Tadzio and Durkheim/Ludwig. This latter theme deserves examination, but first the parallels with *Il Lavoro* justify brief mention.

Il Lavoro, based on a short story by De Maupassant, concerns a couple about to separate following the wife's discovery that her husband has been unfaithful to her. Her source of information is the scandal sheets, which bruit the liaisons he has had with expensive callgirls, to some of whom he has paid large fees.

Pupe is as shocked by her errant husband's expenditure as she is by his loose morals. Although G. Nowell-Smith has described the film as being realized largely in terms of dialogue there is in fact an agile use of mobile long takes within the apartment, during which Pupe pouts and sulks, rants at her uncontrite husband, and talks to her rich daddy in Germany on the telephone. Daddy is withdrawing her allowance, and since Pupe is leaving her husband, she must get a job. She proceeds to assess her own appeal as a potential whore. 'How many times have we made love since we've been married?' she asks. 'Thirteen months. Say, one-hundred and fifty times. At rates you pay your girls you owe me sixty million lire. I could start a boutique, open an antique shop or art gallery.'

Her husband, by now aroused, attempts to bed her. Pupe insists on payment first. He writes the cheque. The phone rings. Pupe blocks the call. 'I've found a job. I'm working.' Her husband returns with the cheque. Pupe cries. The crying, however, seems like a breakthrough in a woman of hitherto marionette-like motivations.

What is particularly interesting, in comparing *Il Lavoro* to the later 'chamber' film, *Conversation Piece*, is the difference in the manner of Visconti's *mise en scène*. The fluency of the camera is reduced in the full-length film, as though Visconti's chair-bound eyeline on the sets proved constraining to him, always in the past so energetic a director. James Price, reviewing the film in *Sight and Sound* (Spring 1976), commented that 'the falsity and pretension of much of the dialogue is severely dispiriting'. It seems likely, however, that the wooden feeling of the lines is due less to a poor script (supplied, after all, by Visconti's regular scenarist) than to Visconti's diminished control over the shooting of his scenes.

In terms of theme rather than *mise en scène*, though, the parallel between the two chamber films grows stronger. Just as Pupe is essentially a spoilt rich girl masquerading as an outraged moralist, so Konrad is a spoilt rich kid (by proxy, helped by Bianca's income) who likes to believe that he is a dissident – the script hints somewhat coyly at his student activities on the barricades in West Germany back in the late Sixties. It is Konrad's sense of pique, as it was Pupe's in *Il Lavoro*, which provides the dramatic triggers for Medioli's story.

The central 'exasperation' in *Conversation Piece* exists in the love-hate relationship of Konrad and Bianca. 'Pull out your phone,' Konrad advises the dazed Professor at one point. 'She'll break your balls.' But the central rapport is that between Konrad and The Professor. The Professor finds the young man's brutal candour fascinating. 'Strange house,' Konrad confesses. 'I don't like it. But it fascinates me.'

231

Claudia Marsani and Burt Lancaster in *Conversation Piece*

The Professor sees Konrad as distinct from the Brumonti family. Konrad, like himself, is socially isolated, and his hostility towards people like the Brumontis has something deeply defensive about it. 'I'm not interested in people like that,' he shouts, 'who aren't in control of their own destiny.' But in fact it is people like the Brumontis, as Medioli and Visconti know only too well, who may well represent the destiny of Italy, as of other Western European nations. Against their monocentric determination to grab all they can, thwarted idealists like The Professor and Konrad have little power of resistance except to retreat into self-chosen ghettos – aesthetics for The Professor, terrorism and hoodlum enterprises for Konrad.

For all the stiffness of the *mise en scène*, and in some of the writing – notably in the vignettes of the young people – *Conversation Piece* manages to touch upon a pointed social theme with considerable acumen. Against his will at first, then later more willingly, the Professor finds himself drawn into the lives of his new tenants, and obliged to align his 'meditative-acquisitive' disposition with their 'frantic-acquisitive' obsessions. When they go away, he finds himself acting as a voluntary service department, taking in building materials, even obliged to choose the bathroom tiles for Bianca. His reward? When the Brumontis return from another jet-setting expedition, they bring a present for him, an exotic talking bird with some resemblance to a magpie. What more pertinent symbol could they offer to this refined kleptomaniac, for it was his acquisitive streak which enabled the family to gain access to his building.

Meanwhile, the rapport between Konrad and The Professor is strengthened as The Professor begins to transfer his affection from his art objects towards Konrad. When Konrad is beaten up by mobsters, it is The Professor who tends his cuts and suggests calling the police. Konrad knows better. 'The police don't give a damn,' he says. 'I might get on some file.' This total absence of faith in, or respect for, the forces of law and order is one of the most accurate character details in the Medioli/Visconti conception of Konrad. The Professor, in his turn, confides that, like Konrad, he once lost faith in the integrity of the institutions, which is why he gave up his profession – science – he saw it as 'enslaving people'. Konrad, however, drops off to sleep. Other people's crises of conscience do not concern him.

In some curious way, however, The Professor seems to need the Brumontis. Once the new apartment is renovated, Bianca invites The Professor to dine. For Visconti, this scene is the most potent in the film. He commented, 'This scene allows the characters to face the most atrocious truths.' These 'truths' include the fact that Signor Brumonti, the right-wing industrialist, has just left Rome following his complicity in a Fascist plot, which Konrad has denounced.

Perhaps more tragic, however, is The Professor's appalled reaction to this bitter dinner-table conflict – Stefan, Bianca's son, is at loggerheads with Konrad, politically and emotionally. The Professor cries, rattled at the breakdown in basic civility as Konrad storms out, 'Everything has turned out far worse than I imagined. The family has turned out to be – a family.' A note arrives from Konrad, 'I hope I'm wrong but I don't think we'll ever see each other again. Your son, Konrad.' An explosion follows, in the upstairs apartment. The Professor drags Konrad from the gas-filled apartment. Konrad is dead.

In the final scene Bianca and her daughter visit The Professor as he lies, perhaps terminally ill, on his hospital bed. 'Grief is as precarious as everything else,' he whispers. The film ends, as it began, on the electroencephalograph tape registering his feeble brain waves. The Professor, too, is now 'enslaved' by the 'caring' technology from which so long ago he decided he had to escape.

Conversation Piece, although it shows the milieu of the new rich, is also a reinvocation of the world of Visconti's youth. The apartment of The Professor, for instance, is modelled on an apartment decor of Visconti's own family, while the Professor's collecting mania, despite Visconti's denial of any autobiographical resemblance on this point, would seem to resemble the director's own passion for collecting art objects. Pia Soli, in the article 'Tribute to Childhood' (*Lumière du Cinéma* No 1, February 1977), observed of Visconti's apartment, 'It was filled with collections of *objets d'art*, pictures, flowers and especially roses – the *grey* variety which he had succeeded in growing in his villa in Ischia.'

The Brumonti family, though, are far from mere touchstones for some

233

Conversation Piece

personal process of recall on the part of Visconti. They are of a world which is all too palpable today. Pushy, strident, shrewd, insensitive and earthy, they are a threat to the Professor's perilously hoarded equilibrium. But they are, equally, representatives of a class of survivors. The Professor, no Sicilian aristocrat like Prince Salina, cannot fend them off as Salina did Don Calogero, safe within the supreme distance of a feudal privilege, however residual. The Professor is a retired technocrat, who renounced his career because he did not like the socio-political implications of applied technology. This shift of alienation is significant for Visconti's development as a social commentator of the mid-Seventies. No longer is he probing social pessimism and solitude in the rarefied (and culturally approved) context of Lampedusa's Sicily, nor in terms of Gramsci's insights into social groupings within the Fascist era. Visconti now examines the problems of acute social alienation as it exists for a man who comes from a broadly based social group rather than a narrow one.

In this film, as in *Il Lavoro* and *Bellissima*, he relinquishes the epic and tragic perspective for the miniaturist focus. *Conversation Piece* touches upon that engaging human dilemma which he has best summed up himself, in the question he asked in 1976 (*Cinema Papers*) – 'Isn't a man alone ... an interesting figure? We want solitude and as soon as we have it completely, we want people around us.'

With his final film, *L'Innocente* (*The Innocent*) of 1976, Visconti returned to the theme of destructive sexual obsession which had been central to his first feature film, *L'Ossessione*, of 1942, and had been recurrently important to him as a means of focusing the conflicts and joys of human relations. The film was based on a novel by D'Annunzio, a somewhat curious choice of material, so it was felt, given the author's association with political ideas which influenced Fascist thinking. Visconti himself, however, had long been an admirer of D'Annunzio 'as a poet and writer' while being much less drawn to his polemical aspects as a salon Nietzsche. Above all, perhaps, the attraction of the novel was the fact that it portrayed upper-class Italian life around the turn of the century, and thus offered Visconti another opportunity to explore the ambiance of his own childhood, as in *Death in Venice*.

The Innocent concerns the tormented marriage of Tullio and Giuliana Hermil. Tullio, a wealthy man about town, is a typically D'Annunzian figure, believing that man is his own god, and free to follow his impulses, whether sexual or martial – his combative relationships with other men are almost as important to the drama as are his dalliances with women and his oppression of Giuliana. Tullio is embroiled in an affair with a supercilious countess named Teresa Raffo, and tells Giuliana, when she questions him about his infatuation, that his love for Giuliana is no longer passionate but merely sisterly. Giuliana, whose self-confidence is initially damaged by Tullio's disclosures, finds the power to react and make crucial choices about her own life.

Giuliana acts. She becomes the lover of an author named Filippo de Gregorio, whom she meets through Tullio's younger brother, Federico, while Tullio is away with Teresa. Tullio, meanwhile, challenges a rival suitor of Teresa's, the Count Egano, to a duel. After the duel, which he has presumably won, he returns to Giuliana in need of emotional support. He no longer needs Teresa, he tells his wife. Giuliana's infidelity is less confused, however. She now loves Filippo, not Tullio, though decency restrains her from boasting about her affair in front of her husband. Tullio soon divines Giuliana's new interest and, in a brief scene with Teresa, who is anxious to consolidate their relationship with a trip to Paris, makes it clear that his affection is waning. He discovers that Giuliana has left the city to spend the summer in his mother's country villa, reneges on his Paris trip with Teresa,

Giancarlo Giannini and Jennifer O'Neill in *The Innocent*

and arrives unexpectedly at his mother's villa.

These rapid elisions, in which we jump from Tullio's challenge to Count Egano into its aftermath, and then from Tullio's decision to accompany his mistress to Paris, into his arrival at the villa, are dramatic echoes of Tullio's jumpy mental state. Equally calculated is the use of colour, particularly in Piero Tosi's costumes for Laura Antonelli (who plays Giuliana). Once Giuliana understands that she is to accompany Tullio on a nostalgic journey to a summer retreat nearby, which Tullio's mother is thinking of letting to an Englishman, she changes from cream to black; and, as she strolls with Tullio in the grounds of the retreat, there is a disconsolate air about her which Visconti's doleful tracking shots serve to underline.

Giuliana allows Tullio to make love to her with a passion he has not shown before – 'You have been wife to me, and sister, but never before lover.' But now she is wise to him; there is always his insistence that a woman should conform to his inner specifications. An absence of confidence, perhaps, impels him to categorize his romantic experience. This defect the Countess Raffo is to expose in merciless terms during her final encounter with Tullio.

Tullio takes Giuliana back to his mother's, jubilant, believing he has retrieved his marriage. His mother (played by Rina Morelli, who died soon after Visconti) tells Tullio that Giuliana is three months pregnant. Tullio weeps secretly while she prattles, unaware that Tullio and Giuliana have not slept together for many months. Again, discreet time lapses impel Tullio towards ruin. From this node point the balance of power within the marriage has shifted. Giuliana refuses absolutely to have an abortion. During Tullio's long absence with Teresa, Giuliana has learnt guile. Having won her right to bear the child, she makes love to Tullio with an open eroticism she never showed before. Tullio tells her weakly that she could be 'free', as he thinks he is. But there is nothing left in him to back up this bombast. Tullio is no longer a hero to Giuliana.

Tullio next talks to Federico, asking him to arrange a meeting with de Gregorio. But his rival is away, in Africa. Tullio also finds Teresa cannot be counted on for support. She is furious with him for deserting her before the projected trip to Paris. 'You got over it quickly,' she says icily. 'It's not good for a woman. There will not be another day for us.'

Then Tullio learns that de Gregorio has died in Africa of a rare tropical disease (another dramatic elision). 'Where are you?' he asks Giuliana, seeing her blankness, knowing that she, too, has heard that Filippo is dead. The baby is due. Tullio tells the doctor that, in the event of difficulty, he must save the mother and sacrifice the child. 'It's a criminal precept,' says the doctor, echoing Giuliana's earlier verdict.

The birth takes place during a thunderstorm. Both mother and child are well. The doctor confides in Tullio's mother, the only member of the family who seems pleased with the birth, 'She has had a strange reaction. She cries.' To the grandmother, however, the baby 'glows like sunlight'. But she tells Federico sadly that neither mother nor 'father' ever see the child, and no baptism is planned. This ceremony is finally organized by Federico and the grandmother.

Federico by now has had enough. He tells Tullio he is going away. For Federico, as for Otto, Ludwig's younger brother, the turmoil of the older brother's inner violence is debilitating. Federico, however, takes steps in time; leaving, he averts his own retreat into melancholia.

The marriage is now a hell. Tullio is continually tormented by thoughts of 'the intruder'. 'I curse the day I accepted him,' he tells Giuliana. 'You love him and pine for the father.' Giuliana denies this; she claims she hates the child for what it has done to their marriage. She even makes advances. But, at midnight mass on Christmas Day, Tullio commits himself irretrievably to a dark road. While the entire family and servants are at the chapel, he puts the baby on a window-ledge outside, as snow falls. He brings the child in only when he hears the party returning after the service. The baby dies that night. Giuliana now lets her true feelings show. She tells him that she only concealed her love for Filippo to save the baby from Tullio.

Tullio now returns to the city. In a bleak scene with Teresa he realizes that she, too, loathes him. In his house (for the first time) she looks around curiously, telling him that the place makes her feel melancholy. She adds that she can see that he is more in love with his wife than ever before; he has never, in her eyes, seemed so pitiable. 'You're vanquished,' she says.

'Your two rivals [Count Egano and Filippo] are victorious, because they're dead. The day comes when we cease to live and start to exist.' Teresa adds slyly that, for a mind at the end of its tether, the consolation of grace can help, but not for Tullio, an atheist.

As she loosens her long black hair, Teresa remarks, the bitterness ceding to a simple sadness, 'You men carry us to the stars, and drag us down. Why can't you love us side by side?' Having delivered what is to be Visconti's final question on the eternal imbalance of human sexual relations, she adds softly, 'I don't love you. You're a monster.'

Tullio is finished. He tells her to stay awake – 'I know how to conclude it.' He steps to the terrace and shoots himself. Teresa hurries away down the drive. Dawn is breaking.

The Innocent is a film about a man who, through his crucial errors of judgement in the way he relates to women, is destroyed by their eventual antipathy towards him. Visconti distils his essential observations on the theme of brittle male pride into his final film, with an elegant and terse control. The saga of collapsing sexual relations has a very different dynamic from the great Italian love stories of the mid-Sixties, however. *L'Avventura*, for example, ends on the image of a woman's hand reaching for an errant male's head – an image of pity. Women cannot forgive Tullio, however. In a sense, one could say that the most delicate theme, buried deep within the

Laura Antonelli and Giancarlo Giannini in *The Innocent*

Didier Haudepin and Laura Antonelli in *The Innocent*

surface tissue of the film, is that of women's inability to forgive. It is a more merciless conclusion than that reached by Antonioni.

Reviewing the film in *Sight and Sound* (Spring 1978), G. Nowell-Smith concluded, 'Teresa's blindness to what is outside the game is deliberate; she sees and turns away. Giuliana is confused and hysterical, but her confusion is a symptom of something not acknowledged ... the film never really resolves itself in one way or the other – which has always been the weakness of Visconti's films, but also their strength.' Yet *The Innocent* strikes home very forcibly through Visconti's economy of narrative handling. Only those moments and encounters between the protagonists which are essential to the drama are included. There is a sense of great purpose behind Visconti's rendering of each scene, all the more remarkable considering the less distinguished control of *mise en scène* demonstrated in the previous movie.

As late as 1979 critics were still pursuing the somewhat fruitless notion of involuted romanticism in Visconti's later cinema. Writing in *The Observer* in December of that year Peter Conrad commented, 'The romanticism of the late films – *The Damned, Death in Venice* and *Ludwig* – thrusts them progressively further into the past ... Beautiful as they are to look at they're bad art because of their decadently uneasy conscience about themselves.

They long for the romantic dream to be true, and yet they punish it by destroying it . . .'

What is significant about Visconti's two films of the mid-Seventies, however, is the degree to which they repudiate such a sweeping verdict. While it is true that Visconti was, latterly, much drawn to decadent themes – 'What interests me is the study of sick society,' he remarked in 1975 (*Avant Scène du Cinéma*) – it is inaccurate to suggest that his identification with degenerate social groups or individuals was either closely involved or romantic in its attachment. He wanted, as he said, to *study* such *milieux* rather than associate with them.

Visconti, indeed, refuted the charge of involution in characteristically oblique manner when he commented, in the mid-Sixties, 'When the mind uses pessimism to dig deep, the will takes strength from it.' In the two films of his terminal years, he achieved the composure of will to be able to study decadence rather than merely choreograph it. At the same time, at the core of his cinema, there remains the ancient, entirely appropriate anger about man's inability ever to apply his great ingenuity to cooperative rather than combative solutions. Antonio's anguish in *La Terra Trema*, when he cries that he started the fishing cooperative for all the villagers not just himself, is echoed directly in Teresa's final despairing question to Tullio – 'why can't men and women go side by side?' Visconti, in 1976, approaching death, knew well enough that her question was unanswerable: but he also knew it was still worth the asking.

Lina Wertmuller has to date made eight feature films, and all of these except one – *I Basilisci* (*The Lizards*), of 1962/3 – were made during the Seventies. They are – *The Seduction of Mimi* ('72), *All Screwed Up* ('73), *Love and Anarchy* ('74), *Swept Away* ('75), *Seven Beauties* ('76), *A Night Full of Rain* ('77) and *Blood Feud* ('78/'9). Of these, only three – *The Lizards*, *Swept Away* and *Seven Beauties* – have achieved a limited release in the UK at the time of writing.

Wertmuller has also worked in Italian television, and her TV films include *Now Let's Talk about Men* ('65), *Rita the Mosquito* (a musical for Rita Pavoni in '66) and *Now Let's Talk about Women* in '67. The last of these TV films, starring Nino Manfredi, was not successful, and there followed a four year hiatus in Wertmuller's career, which ended with her first feature film production in nine years, *The Seduction of Mimi*.

Like Visconti, Wertmuller has also worked in Italian theatre. Two of her plays have been produced, one by Franco Zeffirelli, and in addition she has written for the radio. In the past decade, however, her career has swung decisively towards film-making and, although she has thus far generated little impact in the UK, in the USA, particularly during the mid-Seventies, she has been much feted.

Although Wertmuller's willingness to work across the dramatic media resembles Visconti's enthusiasm for theatre, opera, and (late in the day) television, her closest professional association has been with Fellini, on whose $8\frac{1}{2}$ she worked as assistant director in 1962. When she achieved her own debut film, *The Lizards*, she used a number of Fellini's technicians on the production. The film was produced by Olmi's independent company, was set in the South, and resembled Fellini's *I Vitelloni* in theme, being a study of a bunch of small-town layabouts whose much vaunted *machismo* is mainly hot air and cold feet. The Southern motif emerges most forcibly when one youth leaves for the city, only to return, defeated by the strangeness of metropolitan corruptions; somewhat like Antonio in Bolognini's *Il Bell' Antonio*, he has learnt that he can only survive amid familiar tedium.

Now Let's Talk about Men, the only TV film Wertmuller recalls with much affection, is primarily a comic vehicle for Nino Manfredi, and concerns an academic who is locked out of his apartment, naked, with his shower running indoors and no sign of the janitor. In this vulnerable condition, he fantasises four stories which are designed to demonstrate the many ways in which Italian males exploit their women. In the first sketch, entitled 'A Man of Honour', we see a Roman businessman make two startling discoveries: he has just been bankrupted and his wife confesses that she has been for several years a successful jewel-thief. At first he is shocked, but he quickly adapts to the possibilities, and the story ends as he phones a number of rich contracts to invite them – and their bejewelled wives – to a party.

Candice Bergen and Giancarlo Giannini in *A Night Full of Rain*

241

The second episode, The Knife-thrower, shows a disagreeable old circus performer practising his knife-throwing act on his long-suffering and mutilated wife. This time his myopic aim results not in her injury but her death. This makes him extremely angry with her.

In the third sketch, A Superior Man, we see a crippled husband of odd sexual proclivities discover his fickle wife's attempts to murder him. He entices her into his basement and they grapple in mock death-throes, finally toppling into an open grave he has prepared for this perverted purpose.

The final episode, A Good Man, shows a male supremacist layabout drinking and gambling away his days while his wife toils to bring up the children and perform the chores. He finally goes home to collect his marital rights, and then sleeps while she struggles on with her labours.

Wertmuller has called the film her feminist study, but the bias towards caricature reveals only a nominal commitment to the implications of this ideology. Twelve years after its production, considering the feminist critique of western society in conversation with Gideon Bachmann (*Film Quarterly*, Spring 1977), she made it clear that she believed the feminist demands on the male-orientated economy had become excessive: 'Man ... became a king because he has a slave. And we are trying to take away that last security valve. Imagine what's happening to his frustrations and aggressions now. What will remain of man?'

The Seduction of Mimi takes as its background the curious political events in Catania in the early Seventies, in which the region shifted from Communist to Fascist political affiliations. Wertmuller was intrigued by the shift in political allegiances taking place there and conducted her own researches in the field before she developed her screenplay.

The Seduction of Mimi concerns a Sicilian quarry worker named Mimi Mardocheco, who loses his job when he refuses to cast a vote for the local Mafia candidate. Since his marriage is also going through a bad period – his wife Rosalia snores when he tries to make love to her – he decides to cut his losses in the South and go to Turin, where 'they pay workers well and even the owner of Fiat takes his hat off to them and they are free and respected.' He ignores his friend Pippino's good advice, that the only realistic solution is to stay put and fight for better conditions, and arrives in Turin to find the situation little better. In fact, the man who hires him for a pittance, to work on a building site, bears an uncanny resemblance to the Mafia politician back home – both have three large moles on their cheek.

On Mimi's first day another worker, Nicola, falls from a scaffold and is killed. Mimi is forced to help Tricarico, the boss, dump the body out of town. Nicola, like all the other workers, is a Southerner working without Union protection. Mimi makes it plain that he doesn't like the exploitative conditions, and Tricarico, thinking he has connections with a rival Mafia mob, has him shifted to a factory job, in a metal works. Mimi writes to Rosalia, 'It took me more than a week to come across a native Turinese!' These fond letters he writes home are like a caustic parody of the letters home penned by the factory worker in Olmi's *I Fidanzati*, for while Mimi is conveying affection to his wife he is becoming involved with a fiery girl of anarchist persuasion, Fiore. Mimi also joins the Communist party, and contributes his account of bad working conditions to a Union meeting in which the exploitation of the Southern worker in the North is the main issue. The problem is, no other worker is willing to testify about bad conditions. Mimi realizes why when he sees a Mafia figure in the assembly who looks very like Tricarico – again the three beauty spots. Mimi changes his mind about bearing witness.

As his affair with Fiore progresses he hears from Rosalia, who is also making a new life for herself back home. She has got a job, she is learning to drive, she has had her hair styled. Mimi, too, is undergoing physical as well as mental changes. He wears Fiore's hand-knitted chunky sweaters, and makes love to her in a hairnet, trying to straighten his curls; these domestic scenes are played for their farcical effect by Giancarlo Giannini and Mariangela Melato (who play as a team in the next two Wertmuller movies).

Suddenly Mimi's situation becomes difficult. He inadvertently witnesses a Mafia shoot-out in a cellar, is questioned by the police, and is about to make a statement when he notices that the police Captain also has three moles on his cheek. Again, he decides not to testify. Now he hears from Tricarico that he is to be transferred to Catania, to work in another branch of the company. Although Mimi has not asked for a transfer, he goes anyway; worried that his shady relationship with Fiore will come to the attention of Rosalia or friends of his in town, he smuggles her in, along with the illegitimate son of their union, Mimi II, disguised in a yashmak. Tricarico now approaches him with an offer. If he 'cooperates' with the Mob, things will start going well for him. Tricarico boasts that the town can rapidly be controlled – 'two or three months of bag snatchings, robberies, rapes, a few bombs to be blamed on the anarchists, a little bad publicity about our enemies in Rome and it's done. People will be begging for protection and some order.' To Mimi, such cynical brutality still sounds unattractive.

Now Mimi finds himself caught between conflicting factions in every aspect of his life. At the plant, where he is now a foreman, he is caught between the demands of the workers and his own fear of the bosses, epitomized for him by the malignant Tricarico. Domestically, he is caught between his natural liaison with Fiore and his obligatory relationship with Rosalia, who insists on seeing him now he's back in town. Mimi only resolves this conflict awkwardly: he sees his wife but refuses to make love to her, declaring that he is too feeble. Rumour spreads that Mimi is impotent, which makes him angry. But worse is to follow: he discovers that Rosalia is pregnant.

Here Wertmuller's attitude towards the hypocrisy of the male chauvinist position, so much a comic feature in all her movies, becomes markedly ironic. Mimi, who hears about Rosalia's infidelity from his workmates, goes berserk. 'A Communist doesn't behave like this,' they tell him. 'Communist my ass, I'm a cuckold,' rants Mimi, determined to kill Rosalia with his bare hands.

The interesting development, however, is that Rosalia, when confronted by Mimi, agrees with him. She has dishonoured him therefore deserves to die. What Wertmuller is concerned to point out is that the veneer of industrial culture which has influenced both husband and wife, in their respective search for consumer kudos, all drops away as soon as the basic issues of life are confronted. Mimi has one trump card, as he sees it, however. His liaison with Fiore, and the existence of their love-child, Mimi II, is triumphant proof that he is far from impotent. Rosalia, when she hears the news, does not see it this way. She sets about him in a blind fury. This motif – the woman's jealous fury outstripping the man's – is one Wertmuller much favours; we see it again in the closing scenes of *Swept Away*, in which Gennarino is lambasted by his jealous wife, who has discovered his infidelity with Raffaella.

Mimi's reaction to these multiple harassments is to shift from a 'progressive' to a 'reactionary' stance. He will have no truck with strikes at the plant, declaring, 'Strikes. Disorder. Changes. I've had more than my share. I'll get everything under control.' Part of getting 'everything under control' as he sees it is betraying his workmates; but also he has his revenge on Rosalia's lover, a bulky bureaucrat named Finocchiaro, by seducing the man's wife, an enormous woman with six children of her own, whom he makes pregnant, determining to humiliate Finocchiaro in the town square, before a crowd. This he does, but as always in a Wertmuller drama, the operatic gesture designed to restore the hero's rightful reputation and honour backfires badly. Finocchiaro is duly furious to learn he has been cuckolded, and draws his gun to shoot Mimi. Someone has switched the bullets for blanks. Mimi escapes unhurt, but as Finocchiaro fires, two more distant shots ring out and Finocchiaro falls dead. Someone jams the smoking gun in Mimi's hand. Mimi chases the man he believes to be the killer. A limousine draws up and a cardinal gets out to watch the mêlée. He, too, has three birthmarks on his cheek.

Mimi goes to gaol, where he reads letters sent from Fiore – 'You'll have to learn to become simple, like a child who understands real things,' she advises him. When released he goes to work for Tricarico, his last scruples about abetting the Mafia now dissolved by formidable practical difficulties – waiting for him outside the prison gates are Fiore and Mimi II, Rosalia and her child by Finocchiaro, and Amalia and her six legitimate and one illegitimate child. In the final scene we see Mimi canvassing for Tricarico, and Fiore leaving him. 'Don't leave me. I didn't want to,' Mimi shouts, vainly chasing her through the streets of Sicily. 'I believed in a better world. They're all cousins.'

It is clear from the above résumé that Wertmuller's cinema has a farcical element, which is characterized not only by the comic incidents within scenes but by the risible plots she concocts. One critic has, in fact, compared her to Chaplin – Lilian Gerard, writing in *American Film*, May 1976: 'Like Chaplin, Wertmuller transforms her ideology into absurdities; but, unlike Chaplin, she uses black comedy not as a means of distraction but primarily as a build-up to traumatic endings.'

The comparison with Chaplin seems off-key, however. Where Chaplin thrived on the juxtaposition of tenderness with brutality, Wertmuller thrives on cataloguing human perversities without admitting the relieving factor of tenderness. Wertmuller's cinema is, in fact, a very didactic instrument, designed to express her turbulent ideas about the extinction of the natural resources as a result of over-population. Wertmuller is canny enough, however, to disguise her didactic intentions under the mantle of tragi-farce.

Wertmuller, after years of training in television, commercials, popular play-writing and journalism, has a sharp eye for the enthusiasms and obsessions of the moment. Her films are written very journalistically, with the minimum of description and, despite her use of flashbacks, notably in *Seven Beauties*, little regard for the potential complexities of film structure. Like Shaw, in his plays, she has opinions about society she wants to explain to us, and her plots, characters and themes are all subordinate finally to this pronouncing process. This helps to explain the curious effect her films have on the viewer – provoking, agitating and also exhausting the senses, each scene designed to convey a point about human behaviour, many terminating in a noisy conflict or a quarrel, or often an act of physical violence, with the bare minimum of let-up or release of emotional tension.

Love and Anarchy, features her regular leading players, Giancarlo Giannini and Mariangelo Melato, but this time places them in a period setting, the mid-Thirties. *Love and Anarchy* differs from the preceding and succeeding films in one significant respect: the hero, Tunin. The hero is by no means unsympathetic, as though in conceiving this character Wertmuller briefly declared a moratorium on her general distaste for men of action who are actually puny.

Love and Anarchy shows us how Tunin Soffianti, a Lombardy peasant, is suddenly turned into an anarchist fanatic as a result of seeing a veteran anarchist organizer murdered by Fascist thugs in his own home village. Tunin goes to Paris to make contact with the dead man's bosses, is briefed for a difficult assignment, and turns up in Rome, at the brothel run by Madam Aida, where he is to contact a whore named Salomé, who is also a member of the anarchist organization. Tunin's assignment is to shoot Il Duce during an imminent street parade. Salomé takes him into her protection, telling Madam Aida that he is a country cousin, and offering him not only her body free of charge but also useful information about the Fascist ringleaders who are among her clientèle, in particular one thug named Spatoletti, with whom Tunin has one or two defiant skirmishes during an outing to the country.

Tunin's confusion about the assassination to which he has committed himself is complicated by his growing feelings for another whore working in the brothel, Tripolina, a country girl whose simplicity and vulnerability touches him deeply. When he confesses to Tripolina that he is here to shoot Mussolini she reacts angrily, demanding – 'What the hell is politics to us? What do we care?' Tunin is adamant. He has to go through with the mission. But first, he wants to spend a night with Tripolina, not as a client but as a lover. Tripolina agrees and, after a meal out and a night of bliss, she stays awake while Tunin sleeps, wracked by uncertainty: should she let him sleep on, so that he will miss the appointment on the street with Il Duce, and thereby avoid almost certain death himself at the hands of Fascist thugs?

Salomé comes to wake Tunin. Tripolina forbids her to do it. They quarrel, but Salomé allows herself to be swayed, adding however, 'If you don't wake him he'll hate you for ever,' and posing her own question, 'What would happen if all the mothers and girlfriends stood in the way of soldiers?' To which Tripolina replies, with an asperity which may not be entirely in character for her, but certainly reflects her creator's views on the subject, 'Why not? War isn't the answer. I say, stop this senseless killing.'

Tunin then wakes, and is mortified to find out that he has slept through the street parade and missed his chance of shooting Mussolini. Downstairs some soldiers arrive and he assumes, in his present state of agitation, that they have come for him (whereas they are simply making a routine hygiene inspection). He rushes downstairs with his gun and shoots as many Fascist officials as he can see, before rushing into the street, where he is arrested. As he is taken away in a police van, Salomé shouts at his captors, 'It was for you he was doing this, you bastards. For you who are slaves and don't know it. He was young and he had a heart of gold and he wanted to help you pigs.' Tunin, interrogated by Spatoletti in a police cell, refuses to divulge the names of other anarchists in Paris, and is beaten to death inside a sack. Up comes a caption from Malatesta, '... when their extreme gesture will be forgotten, we shall celebrate the ideal which spurred them.' The caption amounts to a disclaimer. Anarchist violence, even when directed against Il Duce, breeds reservations in the minds of reasonable thinking people, Wertmuller implies, yet retrospectively such quixotic acts must be savoured for the smack of selfless gallantry they carry.

John Simon, Wertmuller's most enthusiastic reviewer in the USA, has commented on Salomé's defiant message to the crowd outside the brothel, 'in a couple of pared down sentences made up of a few denuded words, something essential is caught in all its terrible, shining simplicity.' He adds, with regard to the general absence of sympathy for Wertmuller's cinema in her own country, 'In Italy, where the film world is predominantly Marxist, Lina's anarchic Socialism appears to many movie people to be politically irresponsible. But ... she goes from strength to strength, and is already becoming mythic ...' Is Wertmuller's political vision, as it emerges in her films rather than in what she says in interviews, in any sense 'anarchic Socialism', however? There is nothing in *Love and Anarchy* to suggest she identifies with Tunin's despairing gesture in any wholehearted sense, unlike another Italian film-amaker, Mingozzi, whose film *Gli Ultimi Tre Giorni* (*The Last Three Days*) treated the same political situation with less appetite for the possibilities of burlesque. His story is based on the death of a real anarchist, Anteo Zamboni, who was lynched by a mob in Bologna for supposedly firing on Mussolini during a State visit in 1926. Mingozzi wanted, in his film, to recover a neglected hero of anti-fascist history from semi-oblivion. Wertmuller, however, wishes to suggest that Tunin's own hot-headedness leads directly to his destruction.

In 1973 Wertmuller made *All Screwed Up*, in which she pursues her main obsession, over-population, in very direct terms, depicting the travails of a young couple in Milan. Marluccia bears twins soon after her marriage to Sante, a Southerner only just arrived in the city, and follows the double birth with quintuplets. The story seems to take up Mimi's struggle where *The Seduction of Mimi* ended. Sante, to keep his large brood, is forced to work night and day, even resorting to male prostitution. There is an effective scene in which he lets himself be picked up by a cruising woman, and discovers she is also 'on the game'.

Sexual degradation, a theme which recurs through Wertmuller's films, is examined across a range of characters in *All Screwed Up*, as it was in Wertmuller's TV film of 1965, *Now Let's Talk about Men*. Other characters besides Sante are forced into prostitution to make ends meet, while Marluccia eventually resorts to abortion clinics to solve her apparently unlimited fecundity, changing her mind only at the last moment, when she sees the grey faces of other patients.

Another woman, Adelina, is presented less sympathetically. She is a consumer-crazed social climber who refuses to marry her boyfriend, Carletto, because he is not sufficiently affluent. Commenting on the film in their study of Wertmuller, *The Parables of Lina Wertmuller* (Paulist Press, New York 1977), Ernest Ferlita and John R. May point out the relationship between female sexuality and economic forces: 'Related [in Carletto's eyes] to woman as tease are those inaccessible women whom he glimpses from a distance at intervals throughout the film. It is this economic spectrum that makes them inaccessible ... beckoning and forbidding at the same time.' For Wertmuller, as for Fellini, sex becomes a source of additional frustration rather than a release.

The fact, too, that she casts an equally jaundiced eye on her women as well as her men has prompted a number of American feminist reviewers to ask the question, 'When will you make a film in which a woman represents all of us?' To which Wertmuller has replied, 'I have a reserved relationship to feminism. There are many things they say I don't agree with. What does count is that there is a form of social organization which precludes equality between men and women. As long as we continue being organized in family

247

Mariangela Melato and Giancarlo Giannini in *Swept Away*

units there is always one – the woman – who pays the double price ... I firmly believe that the family must go' (*Film Quarterly*, Spring 1977).

Fellini once remarked that 'Lina likes to tell fairy tales', and none of her films reveals this so closely as *Swept Away*, which shows her predilection for the romantic castaways theme – a theme already treated by Ferreri in *Liza* and Bertolucci in *Last Tango in Paris*. Raffaella is a rich bitch whose upper middle-class social milieu is acidly established in the film's opening sequence. Her husband is an industrialist, on a Mediterranean cruise in his yacht with a bunch of colleagues and their wives. This was also the basis of the odyssey which occurred for Antonioni's lovers in *L'Avventura*, we might recall. This sea-bound party, however, is a good deal more ferocious. The difference in ambiance between the two yacht parties is not only a reflection of the different temperaments of the *auteurs*, but also a measure of the social changes which have taken place in Italy. These industrialists, for instance, can all be seen sitting on deck together reading *L'Unità*, the organ of the Italian Communist Party. This glancing detail, and others like it, enable Wertmuller to etch in her distaste for the (then) emphasis in fashionable circles on the Marxist doctrine. Raffaella, like Wertmuller, wants none of this sloganizing, however. She breaks in on a tedious card game below deck one night to sneer at the men, 'You use your Marxism as a screen, to hide behind.'

The bulk of her anger, however, she reserves for a hairy deckhand from Naples, Gennarino (Giancarlo Giannini), who is only on the cruise to earn a few pounds, and has no relish at all for Raffaella's domineering manner. She abuses him, and flaunts herself topless, knowing he is leering at her; she smokes a 'joint' in full view of the deckhands, gets drunk when she likes, and finally pushes Gennarino just too far, when she orders him to take her to join a bathing party along the coast, in a dinghy which Gennarino has little idea how to handle.

Inevitably, the engine fails, they are cast adrift and finally drift in towards an uninhabited island. Raffaella wades ashore looking for a Holiday Inn where she can drink iced coffee and recuperate. Gennarino follows, carrying her bag and cursing.

In a Northern industrial city in Italy, hints Wertmuller, the middle-class woman tends to dominate, having more energy. In rough conditions though, man is king, because stronger. The basic reality of Wertmuller's biological argument has of course proved particularly offensive to American feminist critics. Gina Blumenfeld, for example, wondered if Wertmuller was attempting to say that sado-masochistic relations were inevitable between men and women. Wertmuller fenced, 'In this film, the woman represents the bourgeoisie and the man the Third World ... she represents men and he represents women.'

If such a statement makes Wertmuller's film sound as portentous as a middle-period Tavianis' movie, one should take it with a pinch of salt, for Wertmuller's cardinal characteristic is her canniness at handling interviews with an eye to the mental predispositions of her interviewers. No doubt she is equally canny in her working routine.

Swept Away is, of course, just what Wertmuller denies. It is, to transpose Jonathan Rosenbaum's description of *Last Tango*, 'the *second* intelligent movie about fucking'. There exists, naturally, the usual Shavian undertow, which gives rise to our strong suspicion, as we watch Raffaella and her hirsute deck-hand disporting in hovel, on dune and amid jungle foliage, that our emotions are being subtly manipulated, that the characters' unreality is a prime sign of the author's covert didactic aims. Just as Melato, sated on sea-air, fresh food and uncomplicated coition, manages to inject an intriguing measure of repletion into her unrewarding role as queen bitch, so the script whisks the lovers away from the island. The rescue boat has been sighted and Gennarino is already signalling, although he ought to know, and Raffaella *does* know, that a return to the mainland will end their *entente*.

Raffaella is restored to her husband and his retinue, Gennarino has to face his irate wife and finds that she, like other working class women in Italy in the mid-Seventies, has heard about the nation's divorce reforms. In

Giancarlo Giannini and Mariangela Melato playing socialist man and capitalist woman as they turn apolitical on their desert island – *Swept Away*

the film's final moments, Gennarino faces the sea, shouting that he has been betrayed on three female fronts – by wife, mistress and Mediterranean. His rhetoric reveals Wertmuller's relish for her intractable hero's discomfiture.

It was *Seven Beauties* (1976) which established Wertmuller's rapid ascendancy in the USA. Following the critical controversy which was focused around John Simon's accolades and the equally intense antipathy of Pauline Kael and Bruno Bettelheim, others of her Seventies' movies opened in New York cinemas – at one time four were playing simultaneously. The film depicts the life of a small-time Neapolitan hustler, Pasqualino, known as 'Sette Bellezze', having seven rather plain sisters.

What *Seven Beauties* tries to do is link the Fascism of the streets of pre-war Italy with the atrocities committed in a Nazi camp, via the use of three extended flashbacks. Pre-war Pasqualino is presented as a non card-carrying Fascist sympathizer, too self-concerned to be bothered with the appurtenances of Party membership but sympathetic towards Mussolini's programme, as he admits in conversation with a Socialist internee during a brief scene in a railway station waiting room. Pasqualino is a petty tyrant in his own household, where he appears to rule his seven sisters and elderly mother, all of whom toil all day in the garment factory. Yet Pasqualino, Wertmuller makes plain, is a male supremacist only because his mother wishes him to be.

One of the odd behavioural puzzles of the film hinges on the relationship between Pasqualino and his mother. In the camp, while he is trying to justify his ambition of seducing the camp commandant, he recalls his mother telling him, in his childhood, that a woman needs to be loved.

The emotional intensity of *Seven Beauties* stems partly from this theme, for the camp commandant (Shirley Stöler), a grotesquely fat Hitler fanatic, when confronted by Pasqualino's hang-dog attempts at sexual proposition-ing, shows the same coldness we have seen in Pasqualino, in the acts of brutality he committed before the War – shooting a pimp, Totonno, beating up his sisters and eventually assaulting a patient in a prison hospital to which he has been confined following his trial for murder.

What he discovers, however, is that this woman has no soft core which he can exploit. When Pasqualino musters the strength to 'collude' with Nazism on the most direct level, penetrating the body of the commandant on his hands and knees, he finds himself pushed away. The commandant is unappeased. 'Your will to live disgusts me,' she remarks.

It is the commandant who represents Wertmuller's political vision at its most ambiguous. For Wertmuller herself has often expressed herself quite similarly on the danger of blind procreative urges. She has talked, for example, of 'an anomalous population growth that suffocates like a cancer' (*Film Quarterly*, Spring 1977), while to Ferlita and May she remarked, 'If I have children, I am ... one become five. It's my end and everybody's end. There's no more harmony ... if you increase and multiply like the grains of sand in the desert, you will *have* a desert.'

What the commandant rejects is Pasqualino's animal vitality. What

Giancarlo Giannini as Pasqualino in the mental hospital sequence – *Seven Beauties*

Wertmuller rejects is Pasqualino's vitality *without awareness*. The two positions may be different, but the areas of overlap help to explain why Bettelheim reacted so strongly against the film, and perhaps why Robin Wood referred to Wertmuller's cinema as one of technical brilliance but very great intellectual confusion.

Through her flashback material, Wertmuller shows how Pasqualino's behaviour is born from a desire to make contact with his mother. While he awaits trial for the murder of Totonno, his mother comes to visit him in the cell. Pasqualino's cry as he glimpses her through the bars of his cell goes a long way to explaining his attempt to make contact with the commandant. Having failed to seduce her, he is told to shoot his friend, Francesco (the anarchist, Pedro, having already killed himself by jumping into the latrine tank).

Pasqualino holds the gun at his friend's head. Finally, he pulls the trigger. Now comes the final scene, after the war, as Pasqualino returns to Naples. He finds that his mother and sisters are whores. The house is filled with garish trinkets, and many dolls. To the girl he was fond of before the war, also a whore, he says, 'Now close up shop, let's get married and start making kids. We haven't got much time. I want lots of them, twenty-five, thirty ... we have to become strong because we must defend ourselves. Do you hear them out there?' Pasqualino has now arrived at his own death of soul, Wertmuller is declaring: he is ready to masquerade as a citizen, and his new weaponry will be the wife and children he oppresses.

There are crudities about the juxtapositions Wertmuller devises for *Seven Beauties*. Her confidence in the urgency of her message leads her to exchange narrative finesse for more baroque devices. In the early scene, for example, when Pasqualino and Francesco burst into a remote forest house in Germany, having just baled out of a POW train, Wertmuller seems inspired by *Duck Soup*. Pasqualino erupts upon a Margaret Dumont-like matron, bursting past her to a loaded table, and grubs up handfuls of comestibles just like Groucho. Equally farcical are the moments when Pasqualino tilts his reluctant phallus at the vast windmill of flesh which is the Commandant in all her *déshabillé* grandeur. Again, his posturing is not far short of Groucho's antics with the rotund Ms Dumont.

Just as assertive is Wertmuller's use of set, costume and decor. Here the key role played in her cinema by husband, Enrico Job, the performance-artist, becomes evident. Job, and Giannini, are Wertmuller's long-standing collaborators, having worked closely with her on all the movies of the Seventies. In *Seven Beauties*, Job's influence is strongly discernable, not least in the camp decors, which were assembled in a former slaughterhouse not far from Rome.

Briefly, we should consider the basis of Bettelheim's resistance to the film, since his antipathy is expressed not so much as review but as a piece of polemic designed to counteract what he sees as contemporary apathy regarding the atrocities of the death camps (originally published in *New Yorker* and

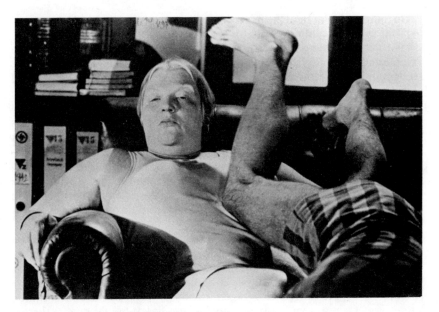

Shirley Stöler and Giancarlo Giannini in *Seven Beauties*

republished in his book *Surviving and other essays*, Thames & Hudson UK 1979).

'Is *Seven Beauties*, then, a film using mocking irony to enrich us?' he asks. 'Or is it entertainment using horrible props to take us more effectively on a ride so engrossing and emotionally exhausting that we are fooled into believing, because of the strength of our feelings, that we have gained in consciousness?' Bettelheim does not deny the seriousness of Wertmuller's intentions but he does state his objections clearly: '... (It) is a somewhat uneasy, indirect, camouflaged – and therefore more dangerous, because more easily accepted and hence more effective – justification for accepting the world that produced concentration camps ... The generally positive response to *Seven Beauties* suggests to me that one generation after the Nuremberg Trials, any manner of accepting Fascism and of surviving under it seems to have become acceptable ... I am discouraged equally by the over-whelmingly uncritical acceptance of this film.' He then cites Michael Wood's remark about Wertmuller's 'great confusion of mind' and adds, astutely, 'Should one assume that those who respond positively to it suffer from a parallel confusion? This may well be the case ... Nothing could be more dangerous than if disappointment with the obvious shortcomings of the free world and life in it should lead to an unconscious fascination with the world of totalitarianism – a fascination that could easily change into a conscious acceptance.'

On the question of Bettelheim's own credentials for comparing Wertmuller's hypothesis about camp-life with his recalled experience of life within one, we should perhaps make it clear that Bettelheim's history does offer him an exceptionally precise perspective from which to focus the film's intentions. He was in Dachau and Buchenwald for a year, in 1938/9, suffered physical violence, and learned to endure by developing a high degree of detachment.

Given Bettelheim's special sensitivity to the documentary realities of camp existence, it seems somewhat cavalier of American critics to dismiss his objections to Wertmuller's film as mere traumatic impedimenta. The fact is, Bettelheim's essay on the film's essentially equivocal aims reveals considerable moral sensitivity. It appears that what he has learned from his wartime experiences is not the bitterness imputed to him by those who dislike his judgement but a greater capacity to resist those choices, in life as in art, which are essentially nihilistic, however adroitly expressed they might appear to be.

Bettelheim concludes his essay in forthright terms – 'our experience did not teach us that life is meaningless, that the world of the living is but a whore-house, that one ought to live by the body's crude claims, disregarding the compulsions of culture. It taught us that, miserable though the world we live in may be, the difference between it and the world of the concentration camps is as great as that between ... death and life. It taught us that there is meaning to life ... a much deeper meaning than we had thought possible before we became survivors.' It may well be that Wertmuller's eventual significance as a film-maker will be that her most controversial movie stimulated a response as finely discriminating to the nuances of cultural responsibility as that offered by Bettelheim.

As a codicil to *Seven Beauties*, we should mention another film which attempted to grapple with the painful reality of concentration camp existence in terms no less ambitious, Andrej Munk's *Passenger*, of 1963, which was left incomplete at the time of his death and subsequently assembled from his partial footage, notes and stills by Witold Lesiewicz. Set in Auschwitz, the story concerns the strange relationship, part hostility part complicity, which develops between a *kapo*, Lisa, and a Jewish prisoner in her section of the camp, Marta. Like Wertmuller, Munk intended to make use of a multiple time structure, setting part of the film in the present, on board a Transatlantic liner. Lisa, now married to an American, is returning to Europe for the first time since the War when she happens to see Marta on a lower deck. At Southampton Marta disembarks without having spoken to Lisa, or perhaps without even noticing her. But the process of painful recall has been initiated in the former *kapo* all the same.

In the filmed sections of the projected movie, largely those set in the camp (the present-day sequences being composed mainly of stills) Munk shows the reality of camp existence. Like Wertmuller, he casts his power struggle in terms of the curious complicities which are formed. Lisa is attracted to Marta; she envies the girl her simplicity and innocence. Stark visual details, too, all the more realistic for being presented in grainy monochrome, add to the inherent plausibility of Munk's fragments. We see a *kapo* set her alsatian on a group of children, a man die on the electric barrier, gauntlets of naked women running past the guard towers, a camp orchestra rehearsing for a concert with the black smoke billowing from the crematorium chimneys in the background. At the same time, hints of the outside world intrude, via snatched news reports on radio stations signalling from London.

Passenger, Munk's film fragment, has passed little heralded into film history. Wertmuller's *Seven Beauties* has been the centre of considerable cultural contention, in the USA at any rate, if not in Britain, where it has been scathingly dismissed and achieved only sporadic distribution. Yet Munk's fragment has set the pace, as it were, for all holocaust movies which have succeeded it. It remains only to say that of the three Italian entries into this most intractable terrain – Pontecorvo's *Kapo*, Cavani's *The Night Porter*, and *Seven Beauties* – none has shown the contemplative power and depth of irony mustered by Munk in his tragically truncated vision.

Wertmuller's films have shown man in authoritarian situations, not man trying to achieve his social liberation. She has yet to make a film which touches on the topics she likes to talk about. In this crucial sense her

Sophia Loren in *Blood Feud*

cinema remains *undisclosed*. She talks of her love of man but her films display a contempt for him, hence the vital importance of Bettelheim's insight that those who persistently portray autocracies may well be unconsciously attracted towards such totalitarian solutions. In a sense, therefore, for all the critical debate which has focused upon her in the past few years, it can be seen that her cinema remains at a preliminary stage. It has not yet become a medium which puts her in touch with her whole personality; her joys, enthusiasms, fears and limitations are not fully engaged: she remains dependent upon operatic plots, elegant sets and a highly theatrical mode of screen performance.

At the heart of her cinema there seems to be a fear at work: of the end of individualistic culture of the sort she recognizes – assertive, combative and given over to gesture rather than contemplation; of the suffocation of cultural forces in Europe by the population explosion of the Third World countries; of the sheer absence of personal ambition of ordinary people, which she sees as 'imbecilic'. At the core, then, there is nihilism of a particular sort; not the poetic nihilism of Antonioni, who believes in no ideology therefore is free to celebrate whatever he sees which strikes him as poetic, without preconception, but the more agitated nihilism of the essentially autocratic personality, which sees in the general absence of social, political or personal optimism an opportunity to impose a private design upon the ethos of – as she sees it – apathy and resignation.

Wertmuller's two most recent films, *A Night Full of Rain* ('77) and *Blood Feud* ('78), were financed by Warner Brothers as part of a four picture contract. The former depicts a contemporary marriage between an American photographer (Candice Bergen) and her husband, who is a Communist journalist (Giancarlo Giannini). The conflict in the drama presents the wife's attempts to define herself – is she primarily a wife or a self-determining professional woman? The film ends with a graphic illustration of this dilemma: she and her husband quarrel; afterwards he falls asleep with his head in her lap. She wants to go, but emotional ties hold her. In the script it says, 'If you fall in love you risk everything, even shutting yourself up in a house to bake bread while the wind howls ...'

Blood Feud saw the end of the agreement between Wertmuller and Warner Brothers, following the absence of acclaim for the previous film in the USA. Wertmuller, criticizing the company's policy towards the movie in *Variety* (3rd March 1978), commented, 'Before reaching the market they have made their own cuts ... if they don't like it, they don't push it. They have nixed the film.' In her second Warner-financed movie she returns to the Italian South for her setting. The time is 1920–2, as Mussolini gains power. There are strong echoes of *The Seduction of Mimi*, not merely the Sicilian setting but also in plot terms. *Blood Feud* also makes use of the vaudevillian device of the woman with two lovers who gets pregnant, not knowing which man has sired her future child. An international sheen has been applied to the drama, however. Sophia Loren plays Concettina, a fiery widow whose young husband

Giancarlo Giannini, Sophia Loren and Marcello Mastroianni in *Blood Feud*

is murdered in the opening scene by a landowner's agent, Accicatena (Turi Ferro), for his rebellious attitude towards the local *padrone*. Titina vows vengeance but it is not until two returning 'heroes' enlist her cause as their own that the ructions commence. One is Spallone (Mastroianni), back from Rome after a long absence. 'I am a Socialist and a lawyer,' he boasts to Titina, who is unimpressed. 'Double trouble,' she grunts. The other is Nico (Giannini), who returns like Lucky Luciano, having made his way in the USA as a bootlegger and killer. 'This is the greatest day in my life,' he announces, arriving in the village square in his ostentatious little yellow motor car. Triumph rather than nostalgia is the motivation for Nico. He has made good abroad and feels fit to challenge any local braggart.

First Nico takes on Spallone, beating him hands down in a card game, during which Spallone forfeits his entire estate, and then refusing to claim his winnings in a gesture whose subtlety seems highly implausible. Later, however, he affronts a more dangerous adversary, Accicatena himself, now in charge of the local Fascists (in this character, in fact, Wertmuller charts a rise to power very similar to that of Attilio in *1900*). Finally, both Spallone and Nico are shot dead by Accicatena's Fascists at Palermo railway station as they try to flee to America with Titina. As the two men lie dying. Titina

257

hugs them both. Her face streaming tears, she murmurs first to one and then the other, 'The child is yours and I love only you, darling.'

This sarcastic conclusion to Titina's troilist experiment is characteristic of Wertmuller's relish for epitaphs which make moralistic points at the expense of psychological acuity. Elsewhere, too, she provides her heroine with lines of dialogue which express not so much the vengeful anguish of a poor Sicilian widow in the early 1920s but the ire of Santa Lina in Rome and New York at the end of the 1970s. 'I can't belong to anyone,' Titina tells Nico. 'That part of me has gone forever.' Thus, crudely delineated emotionalism vies with crudely delineated political confrontation to produce a work which is essentially journalistic in intent. *Blood Feud*, however, the latest in line of Wertmuller's tabloid parables of sexual violence, impinges less forcibly than her earlier Italian-made movies, not least because it is a tired work, filled with echoes of her own output as well as that of her fellow Italian directors. Like *Luna*, *Blood Feud* substitutes *knowledgability* for inspiration. Spallone, the sad lawyer, returning from Rome, where he has evidently been vanquished by harder minds, bears a powerful resemblance to Bell' Antonio of Bolognini's movie, made almost twenty years earlier (the fact that Mastroianni plays both roles of course enhances this lineage), while the scenes between Spallone and his estate foreman resound with echoes of Visconti's portrayal of Prince Salina and *his* gamekeeper, Don Calogero.

Like *Luna*, the film has a fustian quality, crammed as it is with Wertmuller's own aficionado's awareness of past Italian movies, from which she seems unable to unshackle in fulfilling her own project.

'Cartoon strip trying to be tragedy,' one American critic observed of *Seven Beauties* some years ago. This comment catches acutely the essential pressure which Wertmuller's cinema imposes on her audience. She stretches her dramas between the poles of social comment and vaudeville. In this sense she differs markedly from many of her fellow *auteurs* in Italy, making far greater use of farcical and burlesque devices than any of the film-makers cited in this study, though it should perhaps be added that the directors of domestic market comedies – Dino Risi, Luigi Comencini and Mario Monicelli (whose work is conspicuous only by its absence outside their own country) – have also made use on occasion of quite similar plot and dramatic mechanisms.

Wertmuller, however, is very different from these directors too, for she wants to manoeuvre her plots towards a social message. Her surface accessibility, then, is a camouflage for an oddly cold-hearted didacticism. This is the Shavian element to her thinking, as already noted. Problems are finally solved, Wertmuller seems to believe, by talking about them. She is, perhaps, something like Chaplin, whose cinema Parker Tyler astutely defined as that of the 'child aristocrat'. Like Chaplin, in his talkies, speech for Wertmuller 'becomes an emotional and intellectual vice: the magic key to every problem'.

At this juncture in her career, Wertmuller has temporarily retired from the cinematic lists to rethink her position. Following the termination of her contract with Warner Brothers, she announced that she was going to write a

Giancarlo Giannini with Lina Wertmuller

book. With characteristic braggadoccio she decided it would be 'about the poorest people in the world'. Maybe it would be unfair to suggest that such a project sounds like an attempt to refurbish her 'radical' image, now sadly tarnished as a consequence of her relentless self-exposure, which has revealed all too clearly her strengths – her great energy as an artist – as well as her *idées fixes*.

259

Finally, it seems to me, Wertmuller's visually garish and intellectually strident movies may seem most pertinent to the present era of world cinema for what they reveal *latently* about the current shifts in political and social awareness under way in the western nations. Her apparent apoliticality, for example, may reveal the forms of conservatism now taking shape, for there is a curious anti-institutionalism at work. Curious, because this impatience in Wertmuller for any conventional political solution is not in the least anti-authoritarian, as one might expect. It is surely no accident that in the Seventies, the Age of Me, Wertmuller's cinema of personal pronouncement – the Cinema of Me – has flourished.

8. During the Devolution

Between the cinema-as-dream and the cinema-as-polemic, there is an area of middle ground which, increasingly in the past few years, has become a prominent factor in Italian movies.

Considering the relationship between Italian cinema and society in 1977, Gideon Bachmann observed (*Guardian*, 20th April), 'In a political atmosphere which could almost be considered post-revolutionary, inasmuch as the Fascist threat has now largely been exorcised by public fury, the new divorce and abortion laws, and the power position of the trade unions, the people seek clarity.' This observation is interesting not least because what might have seemed plausible in 1977, with the recent gains of the PCI in national and municipal elections, and the measure of social reforms then arriving on the nation's statute books, is now made less valid by present political and social turmoil.

To take three current illustrations of this confusion in Italy, consider the terrorist trial which ended in June 1980, in which 27 defendants were sentenced to prison, or the recent rash of assassinations of city magistrates, or the allegations of corruption which have been levelled at prominent politicians and industrialists following the scandal in March 1980 over a political slush fund and huge unsecured loans by the nation's chief saving banks.

It is evident, then, that Italy today is hardly in 'a post-revolutionary situation'. Indeed, according to a senior member of the judiciary, Lionello Amadei, Italy is 'in danger of being turned into a police state by politicians who are losing their nerve in the face of terrorism' (*Sunday Times*, 13th January 1980). In a situation of worsening economic and political tension the reforms Bachmann cites – the increased power of the trade unions, abortion and divorce reforms – seem little more than cosmetic.

What is significant about Italy's institutional crises, however, in relation to the present directions being undertaken by the nation's film-makers is this: Italy is more despairing than even the black visions of Pasolini or Petrie.

Thus, the role which Pasolini himself saw as crucial – that of scandalizing the Establishment orders – has become redundant. Film-makers can no longer compete with public affairs when it comes to the realization of social nightmares. This leaves the film-makers of today certain alternatives. Some, like those former *enfants terribles* of the more complacent Sixties, Bellochio, Bertolucci and Cavani, have moved towards the cultural traditions; their recent projects in film, theatre and opera have shown their need to pre-empt some of the hallowed ground of Italian culture.

Others, however, seem to have shifted from a romantic to a more classically focused outlook, not least Franco Brusati, born in Milan in 1922 of Austrian parents. In the Sixties Brusati made three films – *Il Disordine* ('62), *Tenderly* ('68) and *I Tulipani di Haarlem* of 1969/70. The first was an episode film showing the struggles of a young Milanese waiter to advance his prospects; the second was a romance about an Italo-American (George Segal) and his childhood girlfriend; while the third was a study of two lonely adolescents in Belgium. Brusati has also written scripts for Zeffirelli and several theatre plays, of which the best-know are *La Fastidiosa* and *The Rose of the Lake*. In the Seventies, however, his films *Bread and Chocolate* ('73) and *Forget Venice* ('78) have attracted international attention.

Bread and Chocolate links in theme to his first film, *Il Disordine*, and considers the travails of a struggling waiter. *Bread and Chocolate*, however, connects this essentially comic motif with a serious social problem – that of Italian 'guestworkers' in Northern Europe. It is a theme which has been focused closely by Fassbinder, in *Katzelmacher* and *Fear Eats the Soul*, and also by Tinto Brass, whose 1968 *Drop-out* dealt with the problems of an Italian immigrant worker in the UK. Brusati, however, sets his film in Switzerland, where Nino (Nino Manfredi) works in a chic restaurant in Zurich, separated from his wife and family back home.

There is a pointed scene early in the film which reveals Brusati's scripting abilities at their best (he co-wrote with Nino Manfredi). Nino takes a break between shifts, waiting for his 'bus boy', who arrives late for work with a cut face. He has been in a fight with local youths who have insulted Italy, he tells Nino. He cannot understand why Italians are resented in Zurich. Nino explains that the indigenous population of a mere five million finds it difficult to accommodate two millions Italians in its midst. If they weren't here the Swiss would be hard put to manage their economy, retorts the youth. Nino demurs. If the Italian workers went home, the Greeks, Slavs and Turks would fill their role as guest-workers, and they would probably assimilate better, since 'they're not as pretentious'. The scene ends when Nino and his assistant go off duty, Nino to his lonely bedsitter existence, the boy to join his girlfriend, a Swiss.

This scene is effective both informationally and emotionally. Facts about guestworkers are conveyed with precision, terse social diagnosis is included – and the point about the pretentiousness of Italians is surely a telling observation in this context – and yet Brusati still manages to conclude the scene by

Nino Manfredi in *Bread and Chocolate*

characterizing the distinctly different possibilities which exist for the two men: for Nino, emotionally tied to Italy, where his family resides, adjustment to this alien, more affluent society is harder than for the youth, who will probably marry a Swiss girl and assimilate rapidly to the new culture.

Brusati himself, in a recent interview (*Guardian*, 27th October 1979), has commented on the culture clash of Italian workers in Switzerland: '... in Swiss hospitals there are many Italian workers suffering from accidents because they are always given the dirtiest, most dangerous jobs. But there are even more in hospitals suffering from paranoia and mental illness. These people, so viscerally tied to their home towns ... unexpectedly find themselves in a world they simply cannot cope with. They are over-awed by its richness, by its superiority in certain ways. At the same time, they long for home – even if it means poverty. My Nino in the film feels this, but is determined to break the barrier.'

'Breaking the barrier' is crucial to Nino. In the end he avoids the error made by many of his fellow Italians, who fall back too readily on the old atavisms – the songs, the wine, the view from the Bay of Naples. In this sense, Nino is a complex character. It is his ambivalence towards his fellow Italian workers which lends the best scenes their conceptual power and emotional tension. For Nino, it is important to stick it out not merely for the obvious financial reasons (he admits at one point that he earns more in tips in one day in Switzerland than he'd achieve in forty years in Italy – his bitterness is evident in his exaggeration) but also because he wants to resist the national dependency on easy sentimentality. Twice he boards the train home, staring in dismay at the card-playing, accordion-squeezing Neapolitans who fill the compartments. Twice he leaves the train, determined to try again, even after his considerable setbacks.

The setbacks incurred by Nino represent the film's weakest, most schematized element. First, Nino loses his work permit after a complaint has been registered by two citizens who saw him relieving himself in the street. He has already been interrogated by the police on suspicion of the murder of a girl. In fact, the police have already apprehended the killer; they are toying with Nino, who, when asked if he is Italian, replies drily, 'Yes, but nobody's perfect.' The authorities are not in fact shown as hostile so much as indifferent. In this significant respect, the film differs markedly from a RAI TV film made by Carlo Tuzii, *Tutte le mattine Domenica*, made in 1972, which also takes as its central theme the conditions of Italian guestworkers in Switzerland, but treats the theme more operatically – at the end one fugitive Italian is killed in a bar brawl, while the young couple who are the drama's main characters endure appalling harassment from the Swiss officials over housing and custody of their only child. Brusati himself has commented, on this question:

'Yes, I criticize the Swiss, but they are simply another rich country that can be indifferent to others ... it isn't, though, a one-way criticism. How have we Italians the right to point a finger at the Swiss when we are making such a painful spectacle of ourselves?' He adds, however, 'My ferocity about my own country is a measure of my tenderness.' It is exactly this conflict of feelings which epitomizes Nino's own ambivalence in the drama. This coincidence of emotion between the author and his hero is the strength of *Bread and Chocolate*. The somewhat contrived storyline is rendered personal and authentic through Nino's testy yet tender struggle against his own sentimentality and his urge to retreat, that quality which he observes and detests in his fellow Italians abroad.

Nino, having lost his work permit, hides for a day or two with a Greek woman, Elena, who has been his neighbour at the pension (played by Anna Karina). Brusati makes the most of the comic potential of these scenes, showing Nino upstaged by Elena's lonely, musically gifted son, Grigori, who is also confined to the room and fills in the long days, while his mother is out at work, playing Mozart with great finesse on the piano – as Nino discovers ruefully when he tries to teach Grigori his one-fingered version of *Il Rossignol*. Elena, though also an exile, has a Swiss manfriend who happens to be an immigration official. When Nino, in one moment of particularly acute loneliness, tries to persuade her into bed, Elena rebuffs him gently, 'We are both lonely. But it isn't a good enough reason.'

Another setback occurs when Nino runs into a rich Italian tax exile who takes a fancy to him, saying – 'We Italians have to stick together. We're all exiles.' He offers to invest Nino's savings for him. Nino goes with him to a private airfield, where the ex-industrialist meets his children, who are typical products of the jet set; bored, blasé, talking in mid-Atlantic accents, oblivious of their father. In a field by the side of a motorway the tax exile shows Nino a photograph of his wife, saying, 'She despises me for squandering a fortune, for laying off workers, as though she cared about the workers.' He tears up Nino's photograph of *his* wife, tells him he is lucky to be poor, since only a poor man still has drive, and offers him a job at his mansion, as butler. When Nino arrives to start work, however, the tax exile has taken a fatal overdose, and Nino has no idea in which bank the man has deposited his hard-won savings. He shakes the already comatose figure, muttering, 'I'll bring you back to life if I have to kick your face in.' In moments like this Brusati shows Wertmuller's relish for scenes of brutal tragi-comedy.

Nino next seeks out an old friend, now working in a construction gang and living in a workers' barracks. One of the film's most moving scenes ensues, during which Nino, his friend and a third (younger) man perform an old drag routine of theirs which they call 'the three graces', mimicking whores as they prance through the bawdy routine. Abruptly Nino stops the act. 'We have to change things not sing about them,' he comments bitterly. There is no hope for him here. Back he goes to the railway station. There he is accosted by a man selling false papers. There is a job to go with them. Nino hesitates – fatally ...

In the hills, amid a gaggle of chicken huts, Nino meets his new 'employer' – an elderly Italian who breeds and butchers chickens as part of the 'black labour' force. 'Every job is done with verve,' he tells Nino gleefully. 'It's like a party. At night we can sleep with a clear conscience, blessing the madonna. We were offered a house but we didn't want to pay the rent.' Instead, Nino finds, the Italians are living in chicken huts. They have adapted to walking at a stoop, and the children 'talk' in chicken language. As they toast Nino, the new arrival at their bizarre camp, he says, 'One question – we're Italian. Is that enough to make us alike?' But his hosts have found fresh diversions. They are ogling a party of rich young Swiss who are picking wild strawberries in the woods beyond the enclosure. The camera closes on one petulant, blonde creature rejecting a fruit. Nino, crouching at the wire portholes, joins his fellow Italians in an act of communal abasement before these strange deities from the Olympian heights of mid-European affluence. He can go no lower. Or can he?

The next major scene shows Nino in a German-speaking bar. He has dyed his hair blond and is passing himself off as Nordic. On the television a football match is transmitted live. Italy are playing. Nino finds himself unable to suppress his sense of patriotism in this situation. He finally erupts, boasting 'Italy will win seven nil.' The Swiss topers are slow to become hostile; they are too indifferent to become aroused by any fresh display of Southern idiocy. But Nino is finally ejected. He lands in the gutter, his blond dye streaking off in a moment which effectively updates von Aschenbach's melting hair dye during the final stages of his delirium in *Death in Venice*. A Spanish whore has followed Nino out of the bar. She helps him up. She, too, passes herself off as Teutonic, but her blonde hair is a wig. Her pragmatic self-abasement contrasts vividly with Nino's abject prostitution of himself, bringing to mind Nino's earlier assertion that, in crisis situations, 'women are calmer than men.'

Nino again boards the train home. Elena arrives at the station. She looks well-groomed and confident. She has married her Swiss official, she tells him, and has arranged a six month work permit for Nino. He is now too defeated to seize the chance. 'I don't have the heart for it.' Nino departs on the train, banging his head impotently on the mirror. He sleeps, and when he wakes it is to hear the inevitable strains of the accordion, the *canzoni di Napoli*, and the conflicts of card-players. The camera closes on his frightened face, just as the train enters a tunnel. A squeal of brakes. Nino has pulled the cord. He emerges on foot from the dark maw, carrying his jacket and suitcase, returning to face the invisible 'barrier'.

'My aim is to reproduce the anguish in which we're trapped,' Brusati has said (*Cinéma 77* No 220, April 1977), 'under the veil of elegance, levity, irony and comedy ... My distance is a form of fear, not coldness ... I have the air of mistreating those I love and favouring those I detest.' *Bread and Chocolate* reveals this combination of distance and affection, showing the hero in a wide (possibly too wide) variety of social situations and pressures. The script is not always certain how to portray the essential clash of cultures – the chicken coop/strawberry picking scene, for example, would have been more effective if handled less lingeringly – but Manfredi's comic range is more evident than in many of his comedy roles. Nino is both buffoon and political mouthpiece. Sitting in his footbath, wearing his flowered towel, he reduces Elena to hoots of laughter, yet he also voices some severe ideas about the defects of Italian sensibility.

Manfredi, incidentally, has recently played an equally demanding role in Montaldi's *Il Giocattolo* (*The Toy*), in which he plays a harassed Roman citizen obliged to hole up in his apartment following his inadvertent witnessing a terrorist shoot-out in a supermarket. The film shows how the unremitting pressure of civic hysteria nudges ordinary people towards a more defensive, authoritarian stance in their private lives and, during its grim final scenes showing the husband and wife in a state of siege inside the apartment, gunmen outside, deep distrust of one another within and, finally, an execution, resembles a contemporary version of Cavani's *The Night Porter*.

By way of a codicil to *Bread and Chocolate*, we might mention Marco Giordana's film, *Maledetti, Vi Amero* (*Unhappy ones, I would have loved you*), made under the auspices of the Jean Vigo film cooperative in 1980 and shown at the Cannes Festival of that year. The film deals with the return to Italy of a young man named Svitol after several years' absence, and has been described as 'a journey through the collective memory of the gen-

eration that in 1968 was in its twenties'. Svitol's perspective on Italy today –
his former friends, the changed social climate, the growing political disintegration – is linked to his ambiguous relationship with a police inspector.
Svitol's sense of detachment and his realization that, in Italy as it is now,
he has become supernumenary, interestingly counterpoints with Brusati's
portrait of Nino, whose problems of identification with his homeland would
no doubt be similar.

Like *Bread and Chocolate*, the most evident defect of *Forget Venice* is the
Bologninian lushness of *mise en scène*, but the visual bravura is not so deftly
counter-weighted by the realism of the story. Brusati has shifted his area of
attention to the bourgeoisie at home – in this case, in a country villa near Venice.

The drama hinges around a family reunion. Nicky, a middle-aged, homosexual dealer in vintage cars and perhaps an ex-alcoholic, arrives at his sister
Marta's spacious villa one summer day with his young boyfriend, Sandro.
His sister is a former *diva* of opera, a local celebrity. Marta's two nieces,
Claudia and Anna (Mariangela Melato), also live here, and watch the bois-
terous reunion of Nicky and Marta. Nicky carries his sister over the threshold,
slips on a rug, and both end up in a heap, laughing. The fall, however, is
symbolic, like much else in this Chekhovian ambience. The following day,

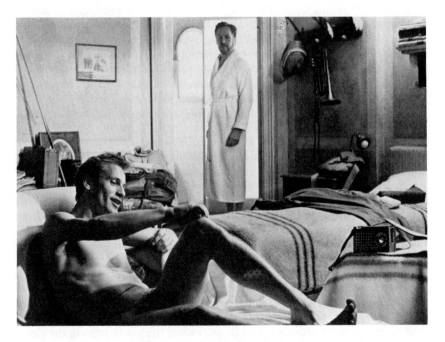

David Pontremoli (foreground) and Erland Josephson in *Forget Venice*

shortly before the group is due to depart on their day trip to Venice, Marta suffers from a fatal stroke. The film ends as the surviving members of the clan vacate the villa, following Marta's death. Grandmother is to stay with a distant cousin – a bemused pharmacist from Milan who is not in the least amused by the faintly bohemian configuration of the household – while Anna and Claudia will stay in Nicky's flat for the time being. But at the last moment Nicky decides to stay on alone in the villa. He bids farewell to his boyfriend, explaining that Marta's death has induced in him a change of values – 'I felt protected by Marta. Now I'm in the front line.' The final shot shows Nicky alone, eating outdoors in the sudden sunshine which follows a fierce storm, dozing in his socks, about to read *Orlando Furioso*. Repletion is the hallmark of this final, satisfying image. But between the arrival of Nicky and the departure of the family much bitterness and self-discovery has flowed. It is in the human observation rather than the somewhat stagey plot that *Forget Venice* achieves its meaning.

The essential theme of the film is the unhappiness of the protagonists, who have failed to mature and are drawn back to the security, and insecurity, of their childhood here in this house. For Nicky, as for Anna and Claudia, the old apple-loft is the room most steeped in memory. Here, Nicky discovered sex, poring over Renaissance nude studies by Giorgione and Botticelli with his lewd-minded pal, Rossino – the flashbacks depicting Nicky's memories are very reminiscent of the hayloft scenes on the Berlinghieri farm in *1900*, in which Olmo and Alfredo also conduct self-initiations into the sexual mysteries during their boyhood. Since a similar use of the apple-loft is central to Marco Vicario's intelligently scripted *La Mogliamente* (*The Wife-Mistress*) of 1978, one is inclined to feel that Italian film-makers today, faced with fatiguing social realities, are in the mood to seek out such havens.

For Anna, whose (more recent) childhood memories appear to be more tormented than her uncle's, the loft is the place where she and her sister took their revenge upon unsatisfactory, neurotic parents, throwing darts at a large formal photograph of mother and father. The mutilated photograph still stands in the loft; like all the other childhood emblems, it serves a somewhat contrived symbolical purpose in the present situation of family crisis and death.

In a later flashback, Anna recalls a terrible childhood scene in which her mother, over-wrought by her husband's infidelities, feigned suicide in front of Anna. Mariangela Melato, doubling the roles of Anna's mother and Anna (as adult), plays the part for all it is worth, almost parodying her familiar termagant roles for Lina Wertmuller.

Nicky's unhappiness is symptomized by his childish relationship with his handsome young boyfriend, Sandro, whom he insists on calling 'Pug'. In a nicely observed early scene, which occurs before they arrive at Marta's villa, they stop to watch a village football match. Nicky urges Sandro to play, and buffs up the paintwork on his Bugatti while the youth joins in the game. When Sandro retires after a fight, Nicky bathes his cuts, proud of his young

Erland Josephson and Hella Petri in *Forget Venice*

warrior's masculinity. The sexual aspects of the relationship are reticently handled, witness the moment when Sandro, hearing Caterina outside their shared bedroom, pushes the double bed into two singles, murmuring to Nicky, 'It's better.'

Claudia and Anna are also, we presume, lovers. They are physically intimate, if emotionally at loggerheads. The repressed, incestuous and fearful situation of these two attractive, not unintelligent young women (Anna is a teacher) seems a trifle contrived; Claudia's role is to play the fussy hen: when Anna feels bad, during Marta's terminal stroke, she fusses around with herb tea, and gets the tray knocked in her face for her trouble. Anna likes to exhibit herself to stray males who might be interested. She exposes herself to a lecherous man who comes to buy apples, and later repeats the performance when alone in the loft during a storm with Sandro. Sandro responds by undressing himself. Anna places a piece of lace on his head as a bridal trousseau. However, when he goes to leave, she begs him to stay. The scene closes as they kneel, holding one another, while the rain fills the culverts in the grounds of the villa.

A key scene, therefore, is the one in which all of them walk through the woods to dine *alfresco* at a local inn, where a wedding party is taking place. Here Marta, once known as 'Golden Throat', so the inn-keeper remembers as he brings the album of press cuttings, takes to the boards again, performing the teasing aria from Carmen. She then dances a brisk tango with Nicky, and collapses, gasping – an early warning signal of the stroke which will floor her on the morrow.

Nicky realizes that the father of the bride at the next table is Rossino, his childhood friend, now an apparently benign patriarch. Rossino has stayed in the area; he is a station master. His just-married daughter is a petulant beauty who is none too pleased about her shotgun wedding to a soldier who has made her pregnant. 'But you liked him when you let him make love to you,' wheedles Rossino, caressing his daughter. This scene impinges forcibly on Nicky, in particular Rossino's advice that he should 'forget mother and father, forget Venice'. He is further disquieted by his glimpse of Rossino making love to his wife in the rest-room. For a moment Nicky watches, fascinated, just as Claudia had been, stumbling across a random fornication in the woods during the opening sequence. As the wedding party leaves, Rossino sends his young son over to Marta's table to present a wedding keepsake. The boy is painfully shy, which prompts the question – given the somewhat unbalanced conduct of his children, can it be that Rossino is not quite the benign patriarch *en famille* that he appears when in public?

This scene is particularly effective because it is wholly cinematic. Too much of the action, set in the villa and making use of somewhat mechanical exits and entrances, seems to derive from a stage conception. The small bird which Caterina releases at the open window is next seen dead on the ground outside. Equally over-insistent is the final image, in which Nicky finds a blue crystal ball (which dates from the days of childhood pantomimes at the villa) and lets it roll away across the lawn, shattering on a tree root. This symbol is particularly obtrusive for coming during a scene which is otherwise nicely handled. Nicky, replete after his meal and his nap, picks up *Orlando Furioso* with the air of a man who can face anything from now on. The recuperative tenor of this closing scene is all the more distinctive for being so rare in Italian cinema of today.

However, for all the lapses in finer visual and dramatic discrimination, *Forget Venice* does consolidate the astringent message Brusati wants to offer to his fellow countrymen. Like Wertmuller, who has also risen to prominence in the late Seventies, particularly in the USA, Brusati makes use of popular forms of story, employing satire, burlesque and knockabout elements to underwrite his social points. Unlike Wertmuller, he has a concise notion of what he wants to say. Italy needs to sing less and look harder at the problems, which derive in the last analysis from internal dissensions.

Above all, Brusati is intent on sealing off the lines of false retreat – the sentimentality, the squabbling, the nefariousness of social relations, the second childhoods. The end, if such tempting illusions are followed, he points out, is grim: chickencoop homes for the working people, apple-lofts for the terrified middle-classes.

Tinto Brass, like Brusati, also began his feature film career in the early Sixties, with *Chi lavora e perduto* (*He who works is lost*) of 1963, since when

The wedding party in *Forget Venice*

he has made nine feature films and a documentary entitled *Il fiume della rivolta* (*The smoke of revolt*) in 1964. The features are: *Il Disco volante* (*The flying disco*) of 1965, *La Mia Signora* (*My woman*) of 1966 – an episode film – *Col Guore in Gola* (*With Guore in Gola*) of 1967, *Nerosubianco* (*Black on white*) of 1968, *L'Urlo* (*The Shout*) of 1969, apparently still unedited, *Drop-out* in 1970, *La Vacanza* in 1971, the last two with Franco Nero and Vanessa Redgrave, *Salon Kitty* in 1976, and *Caligula* in 1977, though the final film has been the subject of much controversy, litigation and harassment, and the existing print of the film, bearing little relation to Brass's movie, has been shown without any director credit.

If Brusati seems at present one of the most ordered of the current wave of Italian film-makers now achieving international recognition, Brass seems superficially his opposite. A self-confessed anarchist, he has admitted, 'I am always on the side of the losers.' Unfortunately for British audiences, none of his films, with the exception of *Salon Kitty* and *Caligula*, is available on the circuits at the time of writing, so our commentary must necessarily remain abbreviated.

During a debate on pornography in Paris, July 1976, in which Pasolini and Alain Resnais also participated, Brass remarked, 'The most banal way of making a modish film on Fascism is to set up a sex intrigue in Nazi Germany.' He might, of course, have been referring obliquely to Visconti's *The Damned* and Cavani's *The Night Porter*, or to his own film, *Salon Kitty*, which depicts the events which take place in Madam Kitty's Berlin brothel during the Second World War. In 1939 she is forced by high-ranking Nazis to sack her girls and employ amateurs who have been selected for their Aryan appeal, both physically and ideologically. Her new premises are wired for surveillance, though Kitty herself does not initially know this (the role is played with considerable *gravitas* by Ingrid Thulin). The new whores become Nazi agents, enticing secrets from their clientèle which are relayed back to the Gestapo. One girl, Margherita, falls in love with an officer, Hans Reiter, who had developed a distaste for the war and deserts.

Hans is subsequently arrested as a traitor by the Nazi officer in charge of the Salon Kitty security operation, Wallenberg, and is hanged from a meat-hook. When he crows about this triumph to Margherita she determines to revenge her lover's death, even while she performs fellatio on Wallenberg. She and Kitty entice him into boasting about his own callous opportunism in a room which is bugged. 'I don't give a damn about National Socialism,' says Wallenberg. 'You're the ones with illusions. It's a way to get you all in our power … we've made criminals out of you, and slaves.'

Kitty and Margherita inform on Wallenberg, who is shot down by senior Nazis in a bath-house in a scene distinctly reminiscent of *Salò*. At this point the war is almost over, and allied Bombers are pulverizing Berlin.

While most of the film is set within the salon itself, and in this respect the parallel with Pasolini's Hall of Orgies is strong, Brass also extends his narrative focus, showing how acts of persecution and ostracism are enacted on the streets of the city. In this respect, the scope of his film is broader, though perhaps less consistently intense than *Salò*, bringing to mind De Sica's final movie, *The Garden of the Finzi-Continis*, of 1972, based on Bassani's novel, which showed the persecution of wealthy Italian Jews in Northern Italy during Mussolini's era.

One element in *Salon Kitty* which differentiates Brass' study from Pasolini's is the attitude of Margherita and Kitty towards the act of sex itself, and its importance in determining balanced human relations. Even though they are forced to witness (in the case of Kitty) and participate in (in the case of Margherita) bizarrely obscene sexual acts, neither woman is shown to be repressed, or in retreat from the positive aspects of carnal experience. These women, Brass suggests, resist the sadistic infantilism of Wallenberg precisely because they are sexually liberated. This is entirely in accordance with Brass' own ideas. 'For me, sex and Fascism are not the same thing,' he has remarked (*Ecran* No 49, July 1976). 'Emotion and romanticism are the true dimensions of anarchy.' If we compare such a position with Pasolini's final conclusion, that 'sex is a form of suicidal self-deception', we can see the point of division which exists between those two provocative *auteurs*.

While *Salò* shows the absolute hold which perversely infantile minds can exert over those in their power, *Salon Kitty* shows the means of resistance available to the victims of such tyrants. Kitty and Margherita fight

Wallenberg with his own weapons – subterfuge and the techniques of surveillance. They are not innocents, like the terrified youths and young girls who grovel at the feet of the 'masters' in the Hall of Orgies. At the end, during the Allied air-raid, Kitty and her company of renegade girls are framed against a background of stained glass windows. It is a mocking visual comment, characteristic of Brass' sarcastic vision of things, but also expressive of his romanticism. Losers can win, the shot proclaims. If the image lacks the terrible, distant tenderness of Pasolini's final shot in *Salò* – the two young Fascist accomplices dancing to swing music in jack boots – it provides instead its own motif of defiant exuberance.

It was Brass' persistent faith in the idea that outsiders can change existing forms of repression, using their deadly weapons – wit, intelligence and will power – which led to the initial problems of his ill-starred film, *Caligula*, of 1976/7. Presented with Gore Vidal's script, Brass proceeded to rewrite it, with the help of Malcolm McDowell (who plays Caligula in the movie). The chief difference in emphasis, according to Brass (*The Times*, 3rd August 1977), was that Vidal wanted to show the rulers as victims of the people, while he wanted to reverse this conception. In the course of preparing the new script, he emulated Fellini's researches during *Satyricon*, plundering Plutarch and Suetonius.

In 1977 Brass commented, 'I believe the ethic of human political reason has failed because we have failed to be guided by strong and precise

Malcolm McDowell in bed with his horse in *Caligula*

sentiments.' Brass felt that Caligula himself indicated this process of intellectual degeneration. To Brass, *Caligula* is structured to show what he calls the 'three logics' – the logic of power (which is immoral), the logic of utopia (which is amoral), and the logic of madness (which is moral). 'Folly – the being different from others – is the only possibly moral way of life,' he insisted. His own view of the producer's antipathy towards his version of the film he expressed concisely in *Screen International* (6th August 1977): 'it didn't reflect the sexual permissiveness of the Seventies and wasn't designed for respectable people and voyeurs.' The same report stresses that the film's producers are excluded from using the existing *edited* version of *Caligula* for any reason, without Brass' permission (my italics). Brass has shown his willingness to fight those who wish to censor his work on previous occasions. During the controversy in Italy over *Salon Kitty*, for instance, during which 76 cuts were imposed upon the print before its release, he tried to instigate legal proceedings against his producer on the grounds that his film was a 'literary work' and therefore inviolable. It seems plausible to suggest, therefore, that, rather than see a mutilated version of *Caligula* distributed bearing his name, he has consigned this version to oblivion, hence the 'additional photography and insertions' credit on the 2½-hour version without his name which has since been shown.

The bastardized *Caligula*, minus Gore Vidal's screenplay and Brass' director credits, opened in London in 1980 in a version containing yet another fourteen minutes of edits. Reviews were predictably hostile, for this bland and rudderless movie lacks all coherence, though the residues of Brass' original abstractions about the absurdity of power still cling to its gunwales. Occasionally, in the sombre photography, the use of deep field composition and the sinuous crane shots, one can discern traces of Brass' distinctive calligraphy, but the Lui/Guccione 'inserts' prove to be joyless soft pornography introduced by a visual device that appears to be a shameless *Casanova* crib – the voyeuristic eye at the fresco spyhole.

Caligula, in this form, is not a film worth more than passing comparison with those Italian epics of the same period it most resembles, *Casanova* or *Salò*. It is filmed erotic pantomime which, in Guccione's hands, seems curiously anxious about those aspects of life towards which its characters feign nonchalance – casual sex, indiscriminate power and, above all, the death instinct – the pornographer's trinity of compulsions, one might say.

Before we consider the careers of Peter Del Monte and Nano Moretti, let us briefly mention Giuseppe Bertolucci, Bernardo's younger brother, who has made three fiction films to date, and a documentary on the filming of *1900*. His first film for RAI in 1972 was entitled *Andare e Venire* (*Round Trip*) and was freely adapted from a short story of Norman Mailer's, *The Notebook*. A young man and woman spend a night together in a railway station. The young man fantasises a series of situations with random travellers, becoming trapped in his own dream logic. Like Brusati, who dusted up his first film idea when he made *Bread and Chocolate*, Giuseppe Bertolucci

Helen Mirren in *Caligula*

returned to the theme of lovers isolated in a railway station in his third feature, originally titled *Oggetti Smarritti* (*Personal Effects*) but recently retitled *An Italian Woman* for its 1980 showing at the Cannes Film Festival. His debut feature for the cinema, however, was *Berlinguer, ti voglio bene* of 1976 (*I love you, Berlinguer*), which established his reputation in his own country but has not yet been released in the UK. The story concerned a Tuscany peasant with an intense admiration for the leader of the PCI. Giuseppe Bertolucci has also, of course, co-written several of his brother's scripts, notably *1900* and *La Luna*.

With *An Italian Woman*, Giuseppe Bertolucci looks likely to impact upon the international film scene. Richard Roud, for one, has praised it in the *Guardian* (12th May 1980), comparing it to the early work of Antonioni – 'the same sobriety, the same simplicity, the same formal and aesthetic pre-occupations ... but these are only references; the film is totally original.' The story shows the events which occur during one day in the life of an Italian woman (Mariangela Melato), who goes to Milan Central Station with her husband (Renato Salvatori) to collect her fractious daughter, who has been staying with the grandmother. Also on the station is her lover, with whom she intends to go to Rome. At the last moment, however, she gives husband, lover and daughter the slip, and enters an apparently aimless fugue which is reminiscent of the condition Wim Wenders depicts in his study of temporary non-attachment, *Alice in the Cities* – the parallel with New German

275

film-making is enhanced by the presence of Bruno Ganz, who starred in Wenders' *The American Friend*.

Ganz plays the strange man whom the woman meets; his mental state is more fragmented than hers, but briefly they join forces, their sense of a common purpose increased by a third party, Sara, a young drug-addicted whore whom they try to help through withdrawal.

Peter Del Monte was born in 1944 of German-Italian parents and graduated from the Centro Sperimentale in 1970 with a film titled *Fuori Campo* (*Off screen*) which was presented at the Cannes, Locarno and Mannheim Festivals of that year. Also in 1970 he made a RAI TV film, *Le Parole a Venire* (*Words to come*), which was shown at the Pesaro Festival, and in 1973, again for RAI, a dramatized documentary, *Le Ultime Lettere di Jacobo Ottis*, based on the eighteenth-century text by Ugo Foscolo. In 1975 he formed a film cooperative and obtained financial aid from Italnoleggio to produce the feature film for which he is best known, *Irene Irene*, starring Alain Cuny.

Irene Irene portrays the chain of events which occur when Guido, an elderly judge, finds himself abruptly deserted by his wife, Irene, after many years of marriage. 'I have always been fascinated by elderly people,' comments Del Monte (*Screen International* 9th April 1977), 'who have reached a point of harmony with life ... my judge in the film belongs to that class. I was interested in seeing what his reaction would be to a very sudden change in his life.'

To Guido, initially, the departure of Irene seems like a mysterious omen. He is disorientated, and, entering a clinic for a period of rest, he meets Alma, a young patient. Guido is drawn to her fragility, and she is attracted to his apparent stability and composure. 'It must take courage to judge a person,' she says. Guido replies, 'Each of us judges others. How can we avoid it, living together?' Alma insists that she judges no one, but as we observe her making a recovery, it is clear that she is capable of hurting those around her (notably, a young man named Heinrich who is fond of her) by simple acts of inattention. Guido, observing this innocent but nevertheless damaging inconsistency, comments, 'Women today believe in being independent but don't know what they mean by this ... my sense of independence is made up of the patient, daily construction of myself ... anyone seeking liberty must conquer himself.'

Guido's relationships with his family are also indicated. To his son, Silvano, Guido remarks, 'Your mother is going through a period of disquiet. It is necessary to be helpful and supportive.' At the same time his daughter-in-law, Emilia, is defined by her mediating role between Guido and Irene – she reads the judge Irene's letters, for example. Irene has returned to her home town, Cividale, knowning that she has not long to live. Guido eventually goes there; he is only reunited with Irene by her coffin. He recalls that, during the war when they were separated, Irene taught here in junior school. He forms a brief bond with a former colleague of Irene's, Nino, who offers him information about that hitherto mysterious interlude in his wife's life. At the end, Guido himself dies, in a railway train. His last letter to Emilia she reads tearfully while she watches the children playing in the yard.

In *Image et Son* No 250, Del Monte commmented on his type of film-making as follows: 'I seek to develop a cinema which is not authoritarian . . . ideological cinema contains a paternalistic view of the public.' What he seeks to do, he declares, is 'to create within a realistic context a more ample and ambiguous reading'. Del Monte's distrust of ideology and rhetoric extends, too, to the field of cinema aesthetics; he remarks, with a lucidity which is characteristic, 'Knowing the cinema does not mean being absolutely cinephilic, but having a moral awareness of cinema's methods.'

In Del Monte's cinema the influence of Antonioni and Olmi is observable: with the former he shares an absence of ideological enthusiasm, while with the latter he shares a distaste for disclosing the sensual areas of experience. His films are direct, modest and penetrating. Given the present climate of production in Italy, Del Monte can expect difficulties in obtaining finance for his kind of films.

Another young film-maker who shares Del Monte's sobriety of vision is Faliero Rosati (born in Pisa in 1942), whose debut feature film, *Morte di un Operatore* (*Death of a Cameraman*) (co-scripted with Franco Ferrini and Valentino Orsini), was realized in 1978. The film is a documentary-style inquiry into the mysterious disappearance of a Swedish photographer, Eric Nold, during the 1967 Arab-Israeli war. Ten years later a journalist attempts to find Nold, covering the same tenuous trail through Berber villages on the fringe of the Tunisian desert. Rosati has observed, '. . . beyond the "human truth" of the man who disappeared, I tried to understand . . . what is contained *within* the film, and indeed within the actual image. Rather than adding any-thing to the image, I tried, as it were, to strip it away . . . until I reached . . . degree zero.'

The 'degree zero' Rosati mentioned can be considered to occur in the film at that moment when the hero arrives in a hovel where Nold must have died. There he photographs himself sitting against a wall in the posture of Nold himself, according to the film footage Nold shot which he carries with him and reruns obsessively at intervals during his odyssey.

At this point, however, he suddenly perceives that he is not alone in the ghost town. Whoever is watching him has no desire to be discovered. He sets up the camera ready to run if 'tripped' by an intruder, and ostentatiously departs in his vehicle. When he returns and runs the footage taken by the camera in his absence he finds a startled child's face gazing into the lens. The film ends as he scrawls *Friend* in Arabic on a wall and settles down to wait. Will the reclusive child appear? Rosati's film brings to mind the deft observation of Kenneth Unwin on the puzzle presented by reclusive existence: 'It's an interesting point in semantics whether you know anything about the real recluse; even whether he exists or not. If you know about him, he's not a recluse: if you don't you can't say.' Such an intangible situation is at the core of *Death of a Cameraman*. While Rosati's film has obvious affinities with *The Passenger* (most strikingly in one desert scene in which the angry hero's vehicle has broken down) the film's preoccupation with the formal properties of film itself links it to *Blow-up*.

Finally, two young film-makers, Maurizio Nichetti and Nano Moretti, both of whom have made some impact in Italy during the past two years with their films *Rataplan* and *Ecce Bombo*. Both are popular film-makers, using slapstick and silent comedy routines to convey their essentially satiric purposes to the audience. Nichetti was born in Milan in 1948 and in 1974 founded the Milan-based theatre cooperative *Quellidigrock* (which if translated as 'those of Grock' makes plain the group's attraction to clowns and clowning). *Rataplan*, of 1979, is a film about the comic antics of a theatre group (the cast consisting of actors from *Quellidigrock*) who take their burlesque routines into the villages and small towns of Northern Italy. It shows the love affairs, the rehearsals, the arrivals and the romances, but essentially it celebrates the life-enhancing rituals of clowning. There is a sharply observed scene in the opening sequence which shows the world to which Nichetti (and *Quellidigrock*) is opposed: the hero, Colombo (Nichetti himself), arrives late for a job interview in a high-rise office block. The scene serves as an exuberant updating on Olmi's equally satirical office interview scene in *Il Posto*.

What Nichetti seeks, in theatre as in film production, is a language beyond words and verbal concepts. 'I know a lot of young people,' he remarks, 'who have stopped talking.' Behind this new stress on silent action which he discerns in young Italians he sees a form of disappointment with rhetorical solutions to political problems.

As Nichetti expresses it, the present dimension of his cinema implies 'a path littered with the corpses of people who carried on talking after they had anything to say. I live in Milan and know people who engage in almost hypermanic activity from dawn to dusk ... convinced as they are that action produces results, while words can only confuse them. My film is a funny film ... but has total respect for anyone who takes genuine independent action. My characters are not passive, bored, sceptical or inconclusive ...'

Equally convinced that what counts is the individual who takes *genuine independent action* is Nano Moretti, who was born in 1953 and made a couple of super-8 mm. films (including *Io sono un autarchico* – *I am an Autarchist* – which was blown up to 16 mm. for Italian distribution in 1976) before making *Ecce Bombo* in 1978. The producer was Mario Gallo, one of Italian Cinema's most creative producers of the last two decades.

The film concerns a bunch of young, middle-class Romans centred around Michele (played by Moretti himself), who hang about in a situation without much hope or challenge, just as Fellini's *vitelloni* of Rimini and Pasolini's street apaches of the Roman *periferia* did twenty years ago. The significant shift, however, is that these youths are bourgeois rather than working class.

The film takes its title from the cry of an old man they see cycling by with a large load on his back, impervious to the world, crying 'Ecce bombo!' (which means, roughly, I am a bomb). This elderly stranger becomes a mysterious symbol for that independence of action which the youths find so difficult to attain in their relations with girls, families and one another.

In terms of its form, the film succeeds on a popular level where the mid-Sixties attempts at non-linear narrative, tried by Bertolucci and others, did not. What such *auteurs* understood intellectually – that film-making must be part of the film – but often failed to coordinate emotionally into their subject matter, Moretti achieves. The film uses alienation devices – 'let's use the fourth take,' characters decide nonchalantly. Alienation effects are used for satirical, deflatory purposes. The style of the film is relaxed, sharp-eyed, and much less in love with itself than its most obvious antecedent, *Partner*, made exactly a decade before, when Bertolucci was the same age as Moretti.

Both films deal with rebelliousness, for both heroes, Jacob and Michele, see themselves as iconoclasts and surrealists. Both are films in which style is the content. *Ecce Bombo*'s great asset, however, is its unpretentiousness and unpredictability. 'Ten years ago they were attacking the system,' one character suggests in Michele's group. 'Red shirts, Black shirts, it's all the same,' they complain. 'Political activists are Martians ... we want to have fun.' *Ecce Bombo* has absorbed the lesson of the Historic Compromise; it spells out the terms of the CPI's capitulation to the demands of economic growth and the imperatives of Fiat using the same texture of dissidence employed in *Partner* and *Cannibals* – student sit-ins, bossy professors and querulous parents – but with an underlying eye for the absurdities.

At the core of the movie, beneath the buffoonery, is a steely diagnosis of the contemporary attrition of faith and energy. For Michele's parents, an act of defiance consists of going out to see a war film instead of staying in to watch television. For the Bombo gang, it amounts to talking about girls instead of going out and finding them. Their specialized concept of contact is epitomized in the scene in which they sit around in Michele's room, phoning girls and holding the receiver over the record deck, on which a Puccini aria is playing. The vicariousness of modern life, the absence of coherent direct action, is the subject of this exuberant, sarcastic movie. Visual and aural references to Italian movies abound, from the gang on the beach at dawn, in the style of *La Dolce Vita*'s ending, to the spectacle of two Bombo gang members munching through a melon in the manner of *Blow-out*. All this, of course, is the opposite of depressing, for it throws the responsibility for his life back squarely onto Michele, who knows it. 'Have you noticed the tunnels in Rome are too narrow for tanks?' he asks his friend Vito. The thought came to mind while he was thinking how frightened he was at the prospect of doing his military service.

Ecce Bombo ends in an empty town square. We see men in cars watching whores, to the sound of music like Nino Rota's. In a cafe an old couple dances stiffly to a jukebox. Cesare sits at a table. 'It's like Fellini,' he mutters.

Nano Moretti in *Ecce Bombo*

The final shot shows a girl alone, in a room, to the music of Puccini. It is entirely characteristic of this intelligent, mordant movie that, when the camera stops rolling, the 'hero' is absent.

'No one does his own thing or helps others,' remarks one of the Bombo group during a conversation. This line is uncannily like Clelia's cry in Antonioni's *Le Amiche*, made a generation earlier. Hearing of the suicide of her friend, Rosetta, she exclaims, 'This empty world! Where no one does anything, yet no one has time to think of others.' In Moretti, the Italian cinema of the Eighties may possess an *auteur* to preserve the integrity of vision of its past masters and perhaps embellish their lucidity with his own distinctively contemporaneous insights.

No account of the period of 'normalization' in Italian cinema would be complete without reference to the work of Roberto Rossellini during the last two decades. From the mid-Sixties, up until his death in June 1977, Rossellini was making films for Italian television. This transition from feature films to dramatized documentary Rossellini himself saw as a conscious choice, remarking to Philip Strick (*Sight and Sound*, Spring 1976): 'I think it would be possible for me to be part of the industry, to be a very successful director, to earn a lot of money, to be respected, important. But I rejected all that, for the sake of adventure. I'm seventy and I don't have a penny. I don't care. I go on, and I have fun.'

It might be mentioned, however that Rossellini's shift to didactic films for television was perhaps stimulated in part by the relative failure of his feature film dramas of the early Sixties. His film of 1961, for example, *Vanina Vanini*, a historical drama set in Papal Rome in the 1830s, was received badly at the Venice Film Festival of that year, where, according to a recent report in *Corriere della Sera* (31st October 1980), 'the critics noted in particular a certain carelessness in the story, a pronounced abandon to rhetoric.'

What distinguishes Rossellini's documentary dramas above all perhaps is not only the great degree of visual control but the thoroughness of his research, the relish he shows for penetrating inaccessible subject matters and making them vivid to an audience. In *The Iron Age*, for instance (published in Edoardo Bruno's monograph, Builzoni 1979), he points that, for Iron Age man, the metal had the power to transform his life, and was therefore treated with reverence. Thus, the hammer and bellows were of symbolical as well as instrumental importance. In Angola, Rossellini explains, the hammer 'was venerated, treated as a leader and fondled as a baby'. While, when he prepared the TV series of 1967. *Man's Struggle to Survive* (directed by his son Renzo), he supervised the reconstruction of many old machines – 'We looked at the original designs in the Vatican and we learned how to make the machine work. We even reconstructed the early clocks for the Medici film ... every step is exciting' (*Time Out*, 15th August 1980).

Following *The Iron Age*, Rossellini continued his series of TV films with *The Rise to Power of Louis XIV* (1966), the five-episode serial *The Acts of The Apostles* (1967/8), *Socrates* (1970), *Blaise Pascal* (1971), *Augustine of Hippo* (1972) and the three-part *The Age of Cosimo De Medici* (1972/73), *Anno Uno* (*Italy Year One*) of 1974, and *The Messiah* (1975). These films, made on relatively low budgets, were shot rapidly, using a master sequence, with zoom and pan shot replacing conventional shooting techniques. The films are a stylistic expression of Rossellini' own philosophy, founded on his sense of the importance of simplicity, curiosity and serenity in executing art works. His much cited dictum, 'Have the courage to go slow,' is evident in every frame.

Luigi Vannucchi as de Gasperi in *Anno Uno*

Take the opening sequence to *Anno Uno*, his study of Alcide De Gasperi, the Christian Democrat premier of the post-war era of democratic reconstruction, in which we watch civilians evacuating a village as an artillery attack begins. 'Italy is a mass of rubble,' one evacuee grieves, as the camera nudges us from Red Cross tent to groups of wartime vagabonds to wrecked buildings, and then on (with a vengeance) to a road lined with bodies during a bomb attack. It is masterly film-making; each shot has its own impact and yet is part of a pattern. The reality of war, expressed in a couple of master-shots, with one moving camera.

Mise en scène aside, though, what do these TV films reveal about Rossellini's vision of society? They show firstly his preoccupation with the potential unity of human experience, particularly in the ethical and intellectual sphere; they show, so to speak, Rossellini's capacity to grow old gracefully.

In *Cosimo De Medici: Part III*, standing in contemplation of a Masaccio canvas Alberti remarks to a priest that 'fine art and transcendence are interwoven'. Sitting in a governmental committee meeting, De Gasperi insists that politicians of all factions in post-war Italy must put the national interest first, and remain aware of Italy's relations with the whole of Europe. Nearly five centuries separate the two men of history, but their vision is similar. Rossellini aligned with such men. Also, as head of the Centro Sperimentale, he influenced a number of student film-makers, and among those who openly acknowledge their debt to Rossellini are Marco Bellochio, Liliana Cavani and Peter Del Monte.

This search for a harmonious force in human affairs is bound to meet resistance, and Rossellini's films show it. At the end of *Anno Uno*, De Gasperi is defeated, and retires in 1953 (dying the following year). The reasons for his defeat are not clearly stated in the movie but, according to John Earle, in his book *Italy in the 1970s* (David and Charles 1975), 'De Gasperi opened himself to criticism for allowing the level of party politics to deteriorate by shutting his eyes to the spread of its less reputable practices.' Among the Left in Italy, *Anno Uno* is viewed as apologist support for the Christian Democrat party. *Anno Uno* has not yet aroused much sympathy of response among British critics either. Philip Strick, for instance, described it as 'endlessly pirouetting around a succession of non-events'. While this reading of the film does not take account of Rossellini's main purpose – to show De Gasperi's essentially statesmanlike qualities – it does contain substance, for Rossellini seems less able to penetrate the political netherworld of sub-committee wrangles than, say, a film-maker like Francesco Rosi. For all that, *Anno Uno* remains consistent with Rossellini's over-riding aim, to seek benign solutions, to transcend, in his didactic films, his own sense that 'society is a continuous failure' and to investigate states of mind, and forms of action, which serve as alternatives to incipient chaos.

It remains to be said that his output for television during the past two decades raises provocative questions. Rossellini himself rebuts the claim that his TV films are too recondite for a popular audience. 'Some of the papers ... claim that my films are beautiful but do not "entertain". They call for entertainment as if it was a social right, like health. But what is "entertainment"? Is it not "fun" to know?' The answer, of course, has to be 'not for everybody'. Pleasure in intellectual discovery is, after all, a measure of cultural self-confidence. Do we discern here some evidence of Rossellini's famous obstinacy, that absence of flexibility to which Ingrid Bergman alludes in her recent autobiography?

Rossellini's sanity of purpose, then, is indisputable but, outside Italy, what are the prospects for his didactic films? One brutal illustration serves to underline the issue: *Anno Uno* opened in London in the same week as *Caligula*. *Anno Uno* played to empty auditoria; *Caligula* didn't.

Caligula and *Anno Uno* represent the two poles of Italian cinema today. The one invites you to retreat from reality, the other to embrace it. Retreat takes no effort, pursuit demands energy. Is there any road between these two choices, the personal struggle towards self-knowledge or the impersonal surrender to sensation? Rossellini was fighting a last-ditch stand for cerebral order and harmony. His social diagnoses remain impeccable, but, in the last analysis, inaccessible. The strange absence in his later TV films of any dimension – apart from the disquisition of already informed minds – suggests, perhaps, a certain disregard for the niceties of his scripts – there are moments when group-dialogues prove embarrassingly stilted.

Luigi Vannucchi in *Anno Uno*

If Italian cinema, in the next decade, is to follow the general trend in Europe and move towards TV co-productions spiced with the occasional 'international' epic such as *Casanova* or *Caligula* then the most appropriate avatar for this new era of 'small is beautiful' would seem to be a film-maker like Fellini, who has adapted to TV film-making in the past decade, making short, personal films for the new sitting-room audience – and this despite the fact that he himself has strong reservations about the erosion of cinema, a collective experience in which all share a dream in silence and darkness.

Orchestra Rehearsal, for example, derives its thrust from a very Rossellinian vision. Certainly Fellini, once Rossellini's script-writer and assistant director, would agree with his former mentor's verdict that 'society is a continuous failure'. *Orchestra Rehearsal* also seeks to show us how civic disaffection may lead us to chaos; it, too, proposes a Rossellinian alternative – the power of art to reassert meaning. But Fellini, a satirist, is willing to show us the weaknesses inherent not only in the problem but also in the potential solution. Thus, his orchestra conductor, the man who eventually re-imposes a kind of uneasy order on the warring musicians, is shown in *all* his facets. He may be a force for harmony, unity and order, like Rossellini's philosopher and statesmen heroes, but he is also all too human and concerned with his own interests – creature comforts and self-promotion.

Dare one suggest, then, that Rossellini's transition from cinema drama to TV didacticism, though marked by good intentions, indicates some loss of vitality? Absence of concern with film form, with the primal art of 'story telling' may seem agreeably 'anti-monumental' to Italian intellectuals, but tends to deter the wider public upon which all popular art forms vitally depend.

The question remains: in an age when not only economies but also cultures are gripped by recession, do Rossellini's films go one way, and the world quite another? Rossellini urges us to the consideration of art and science and political achievement. His Renaissance men find the contemplation of Masaccio a vital form of replenishment. Whereas the orchestra conductor, from his podium, faces a riot. Between civil war on the streets, and past glory in museums, Italian cinema continues, a dance in despair.

Index